MATH FOR PRESCHOOL COUNTING 1-10 AND COLOURING SHAPES

Ages 4 - 5

BY: CHRISTINE MARCIA SMALL

Written: OCTOBER 2019©

COPYRIGHT: CHRISTINE MARCIA SMALL©

THIS BOOK BELONGS TO:

_____

INTRODUCTION

MATH FOR PRESCHOOL COUNTING 1 – 10 AND COLOURING SHAPES is a very interesting, educational and informative book for 4 and 5 year olds. The exercises are full of fun and your child will be having fun while learning. It is also colourful. Your child will be:

- Counting and identifying numerals and shapes
- Identifying colours
- Solving numerals patterns/sequence
- Solving shapes patterns/sequence
- Matching numerals that are the same
- Matching shapes that are the same
- Drawing and colouring shapes
- Counting shapes that are in groups and tell how many

TABLE OF CONTENTS

Let us count	5
Numeral 1	6 - 8
Numeral 2	9 - 16
Numeral 3	17 - 29
Numeral 4	30 - 41
Numeral 5	42 - 53
Numeral 6	54 - 69
Numeral 7	70 - 87
Numeral 8	88 - 104
Numeral 9	105 – 122
Numeral 10	123 – 145

LET US COUNT

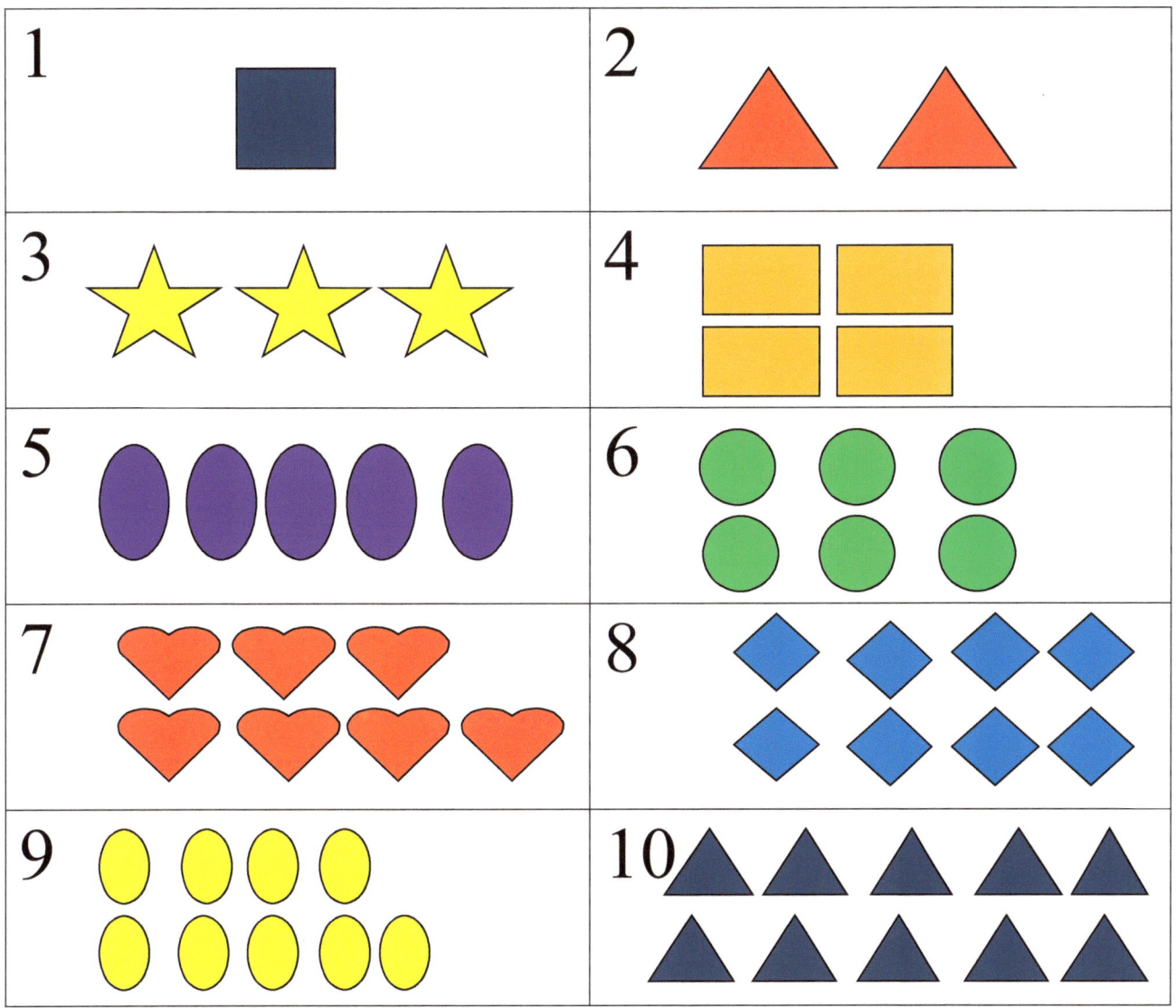

Numeral 1

One blue square

Numeral 1

Colour the square blue.

One square

Draw a square then colour it blue.

MATH FOR PRESCHOOL COUNTING 1 – 10 AND COLOURING SHAPES | 9

 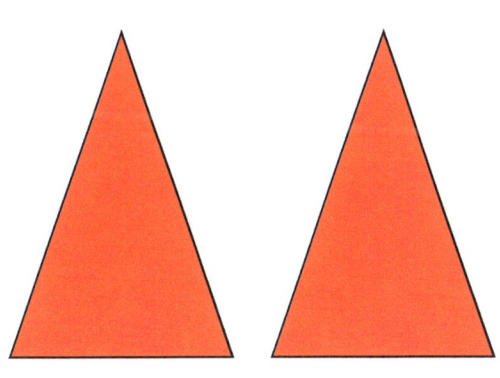

Two red triangles

Numeral 2

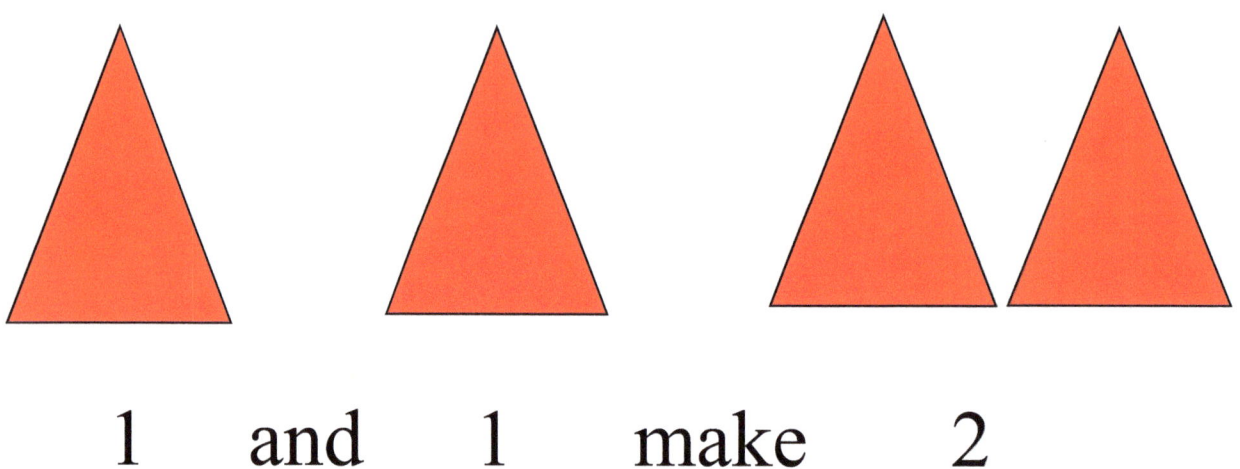

   1   and   1   make   2

Draw triangles to match the numerals below. Colour the triangles.

   1   and   1   make   2

Draw 2 triangles then colour them red.

Numerals 1 and 2

1

2

Draw lines to match the numerals to the correct number of shape or shapes.

1

2

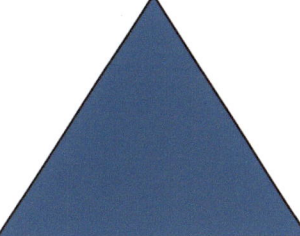

Draw shapes on the lines to match the numerals.

1 _____

2 _____

Numerals 1 and 2

1. Colour 1 triangle blue.

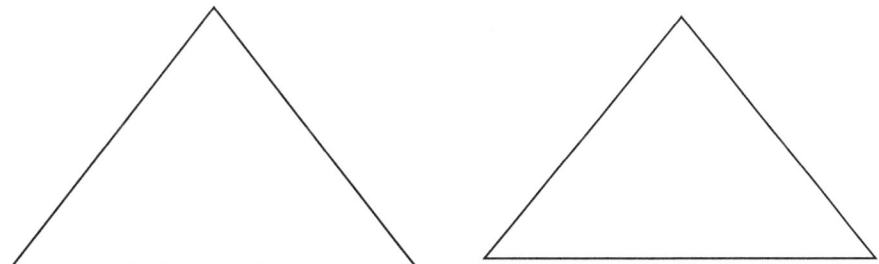

2. Colour 2 squares red.

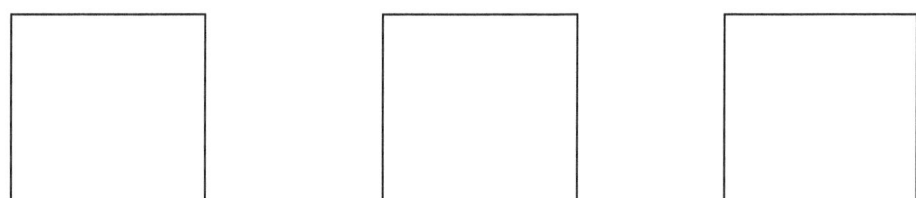

Numeral Patterns

Circle the numeral which comes next.

1 1 2 1   1 2

1 2 1 2   1 2

2 1 2 1   1 2

2 2 1 2   1 2

Numeral 3

Three yellow stars

THREE

1 and 2 make 3

Draw stars on the lines to match the numerals below. Colour the stars.

_____   _____   _____

1 and 2 make 3

THREE

2   and   1 make   3

Draw stars on the lines to match the numerals below. Colour the stars.

_____   _____   _____

2   and   1 make   3

THREE

Count the stars then fill in the blanks below.

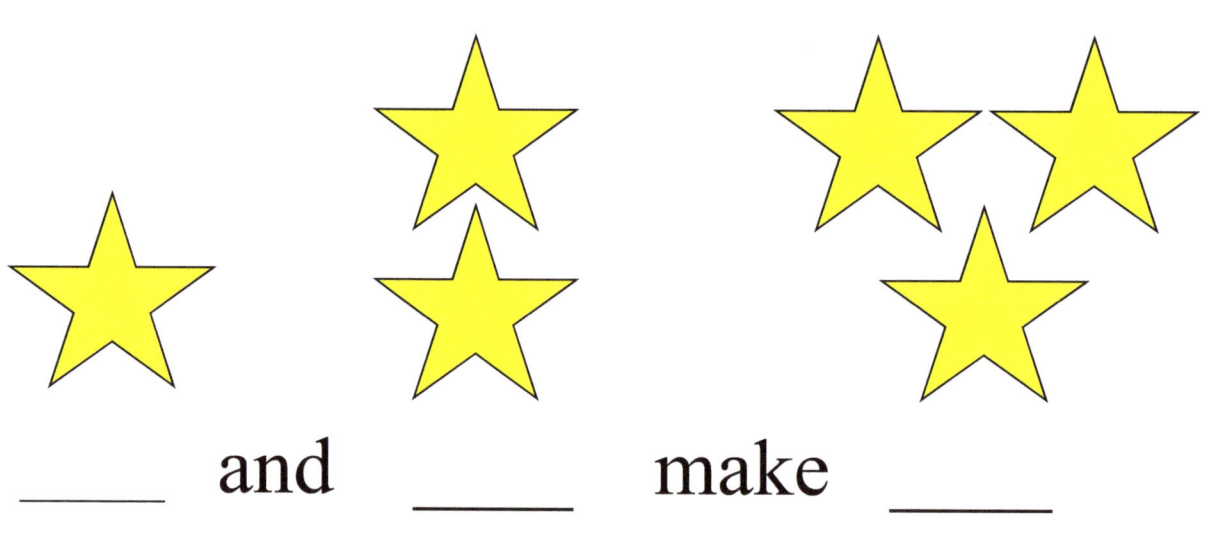

___ and ___ make ___

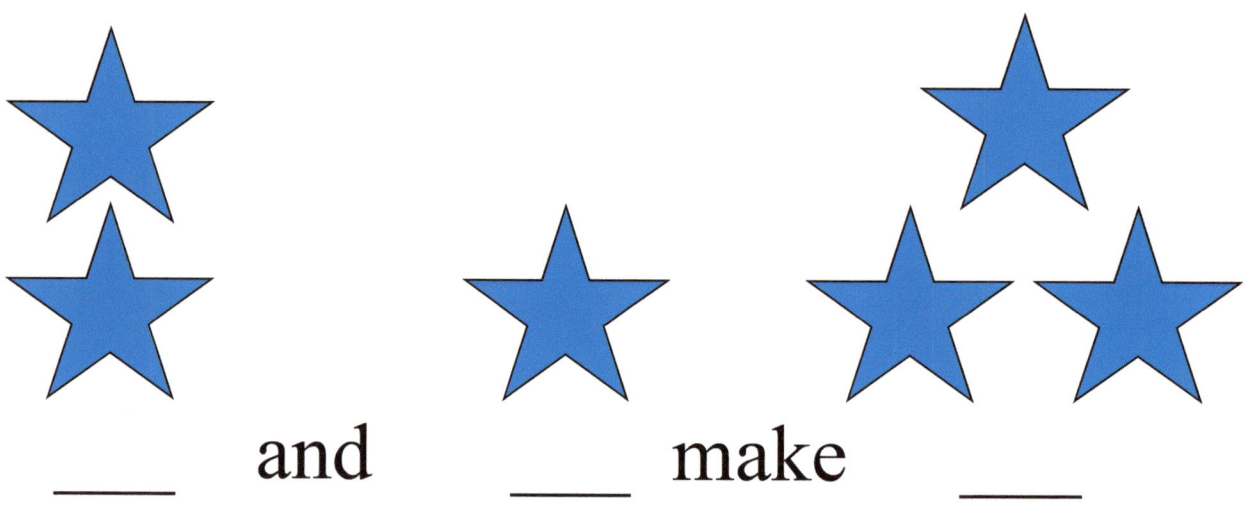

___ and ___ make ___

Draw lines to match the numerals to the correct number of shape or shapes.

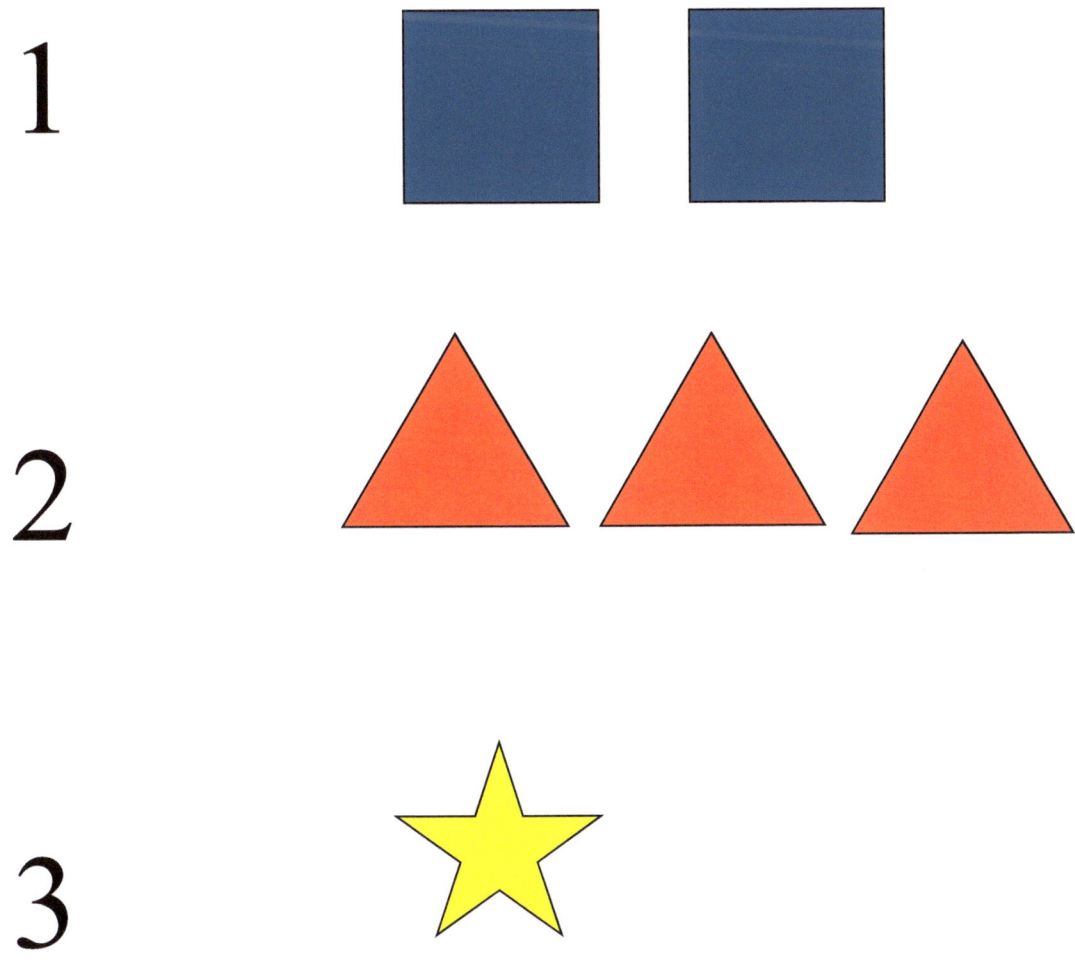

Draw a shape or shapes on the lines to match the numerals.

1 _____

2 _____

3 _____

1. Colour one star red.

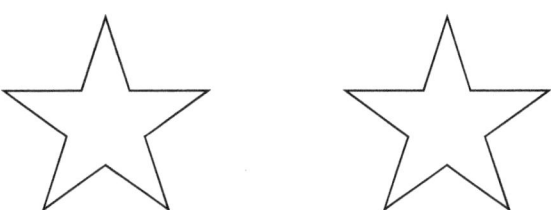

2. Colour two squares yellow.

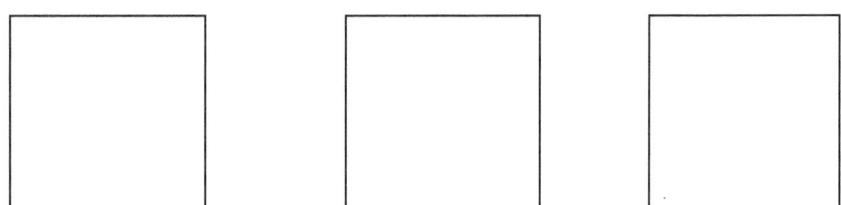

3. Colour three triangles blue.

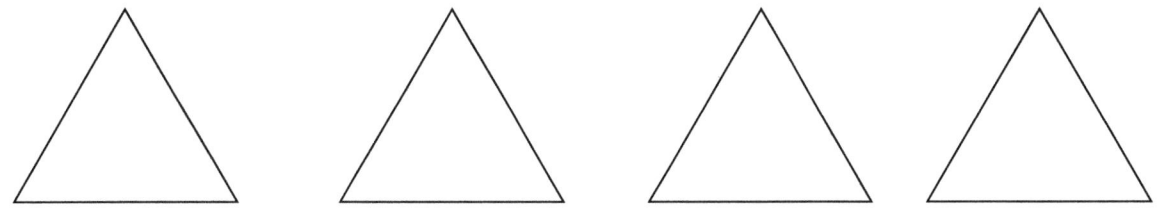

Circle the group of three.

Numeral 3

Draw more shapes on the lines to make groups of three.

1.

2.

Match the numerals that are the same.

1   3

2   2

3   1

Numeral 3

Match the shapes that are the same

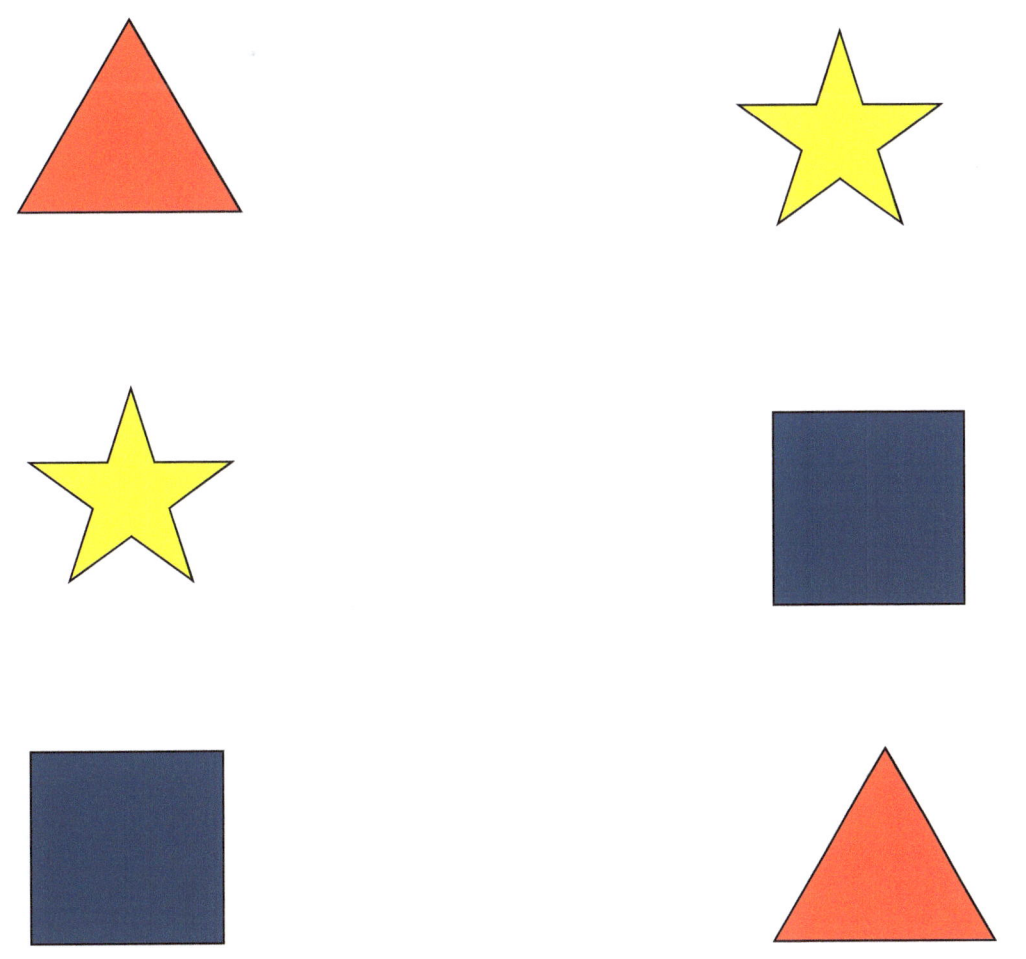

Numeral 3

Draw and colour three stars.

Shapes Patterns

Colour the shape that comes next.

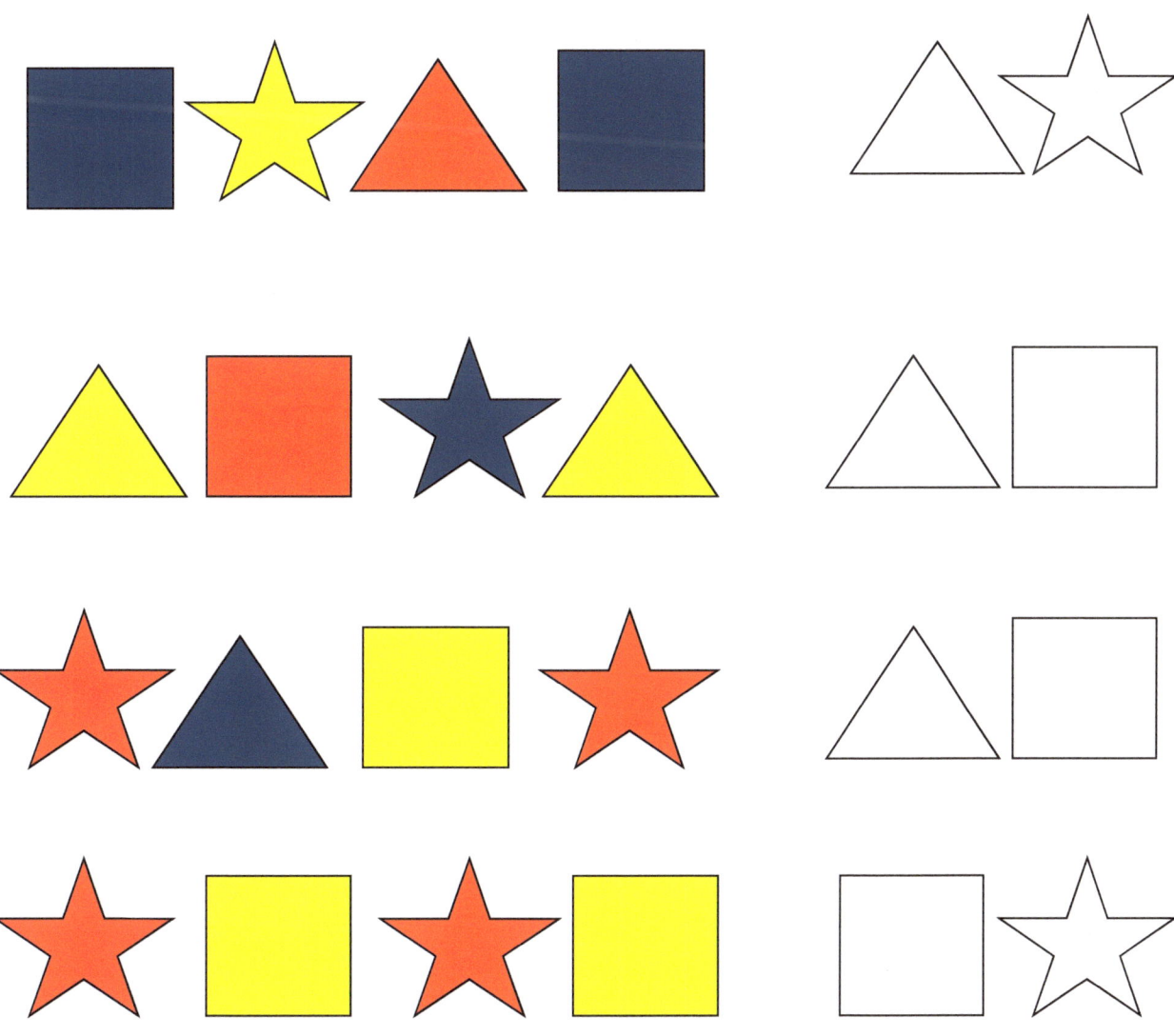

Numeral 4

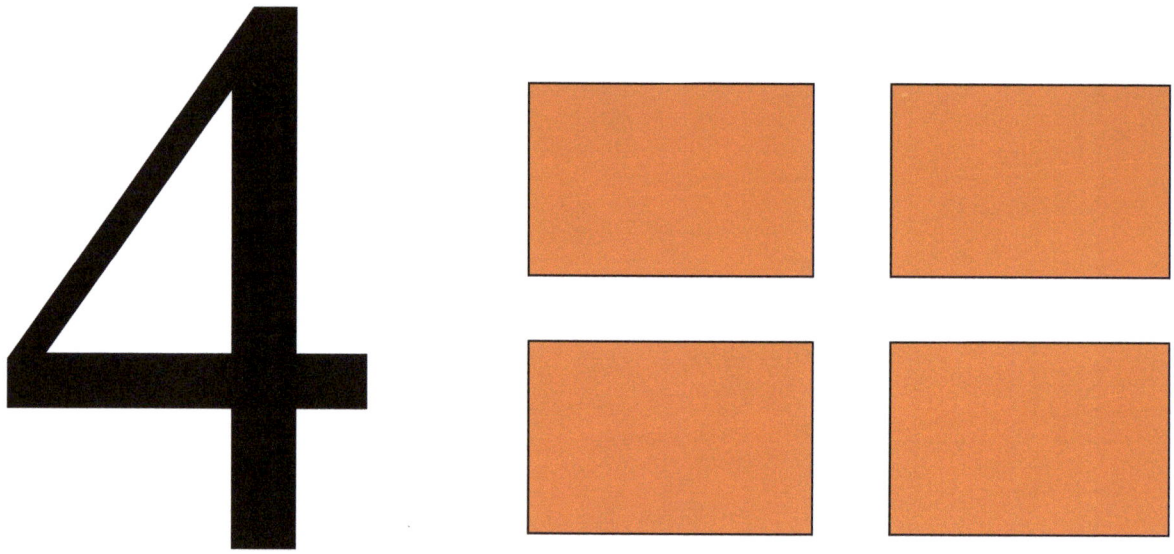

Four rectangles

Four

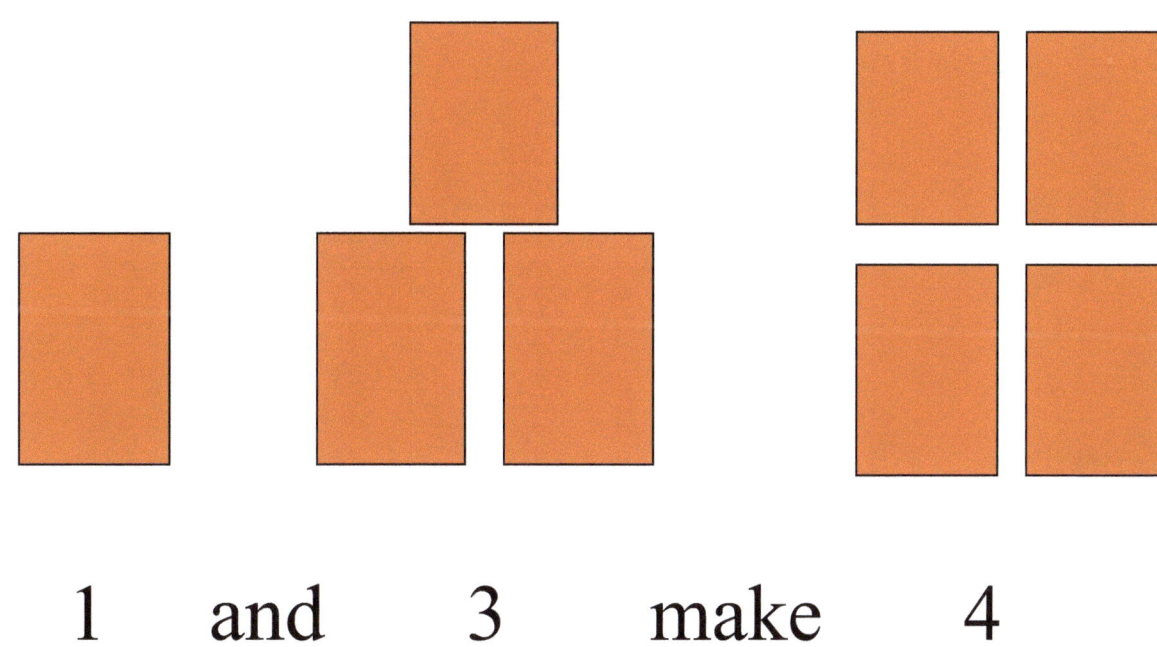

1   and   3   make   4

Draw rectangles on the lines to match the numerals below.

_____   _____   _____

1   and   3   make   4

Four

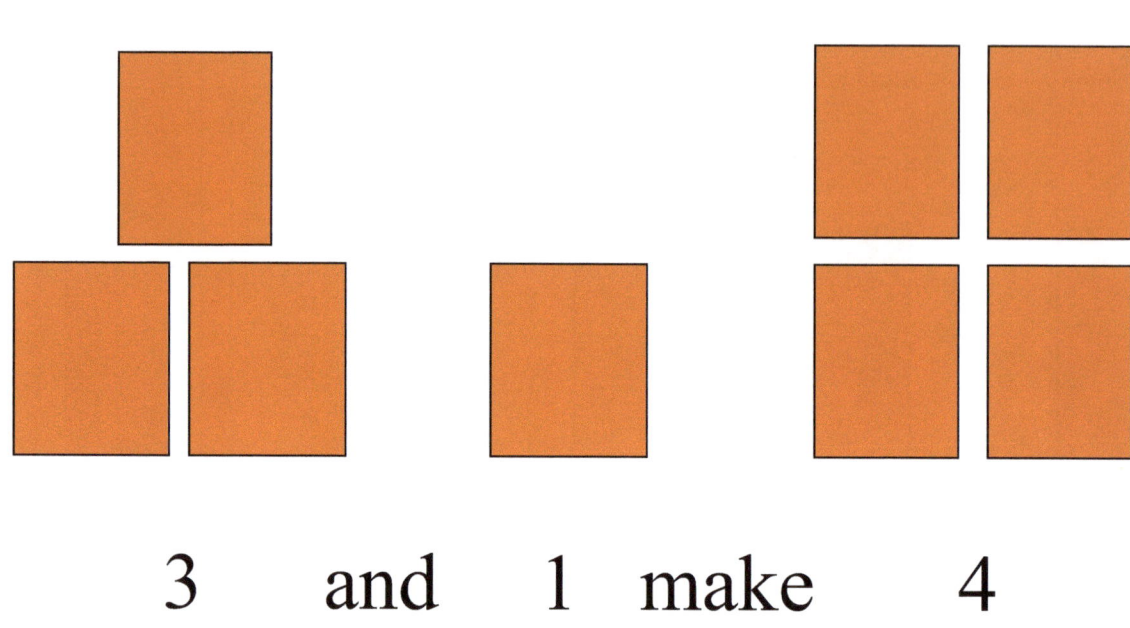

3 and 1 make 4

Draw rectangles on the lines below to match the numerals then colour them.

_____   _____   _____

3 and 1 make 4

Four

2 and 2 make 4

Draw stars on the lines to match the numerals below then colour the stars.

_____   _____   _____

2 and 2 make 4

Numeral 4

Count the rectangles then fill in the blanks below.

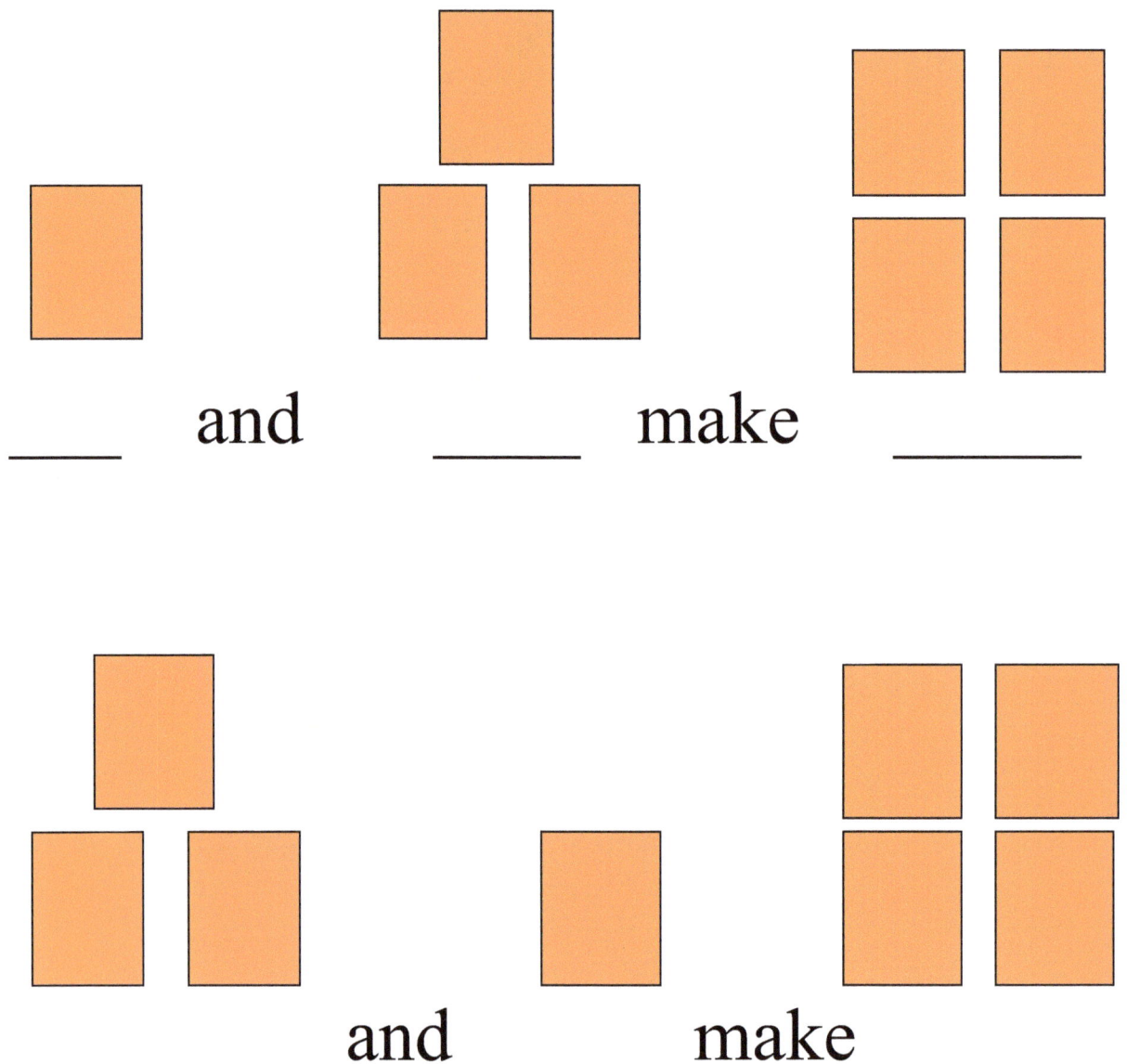

___ and ____ make _____

____ and ___ make ____

Four

Count the triangles then fill in the blanks below.

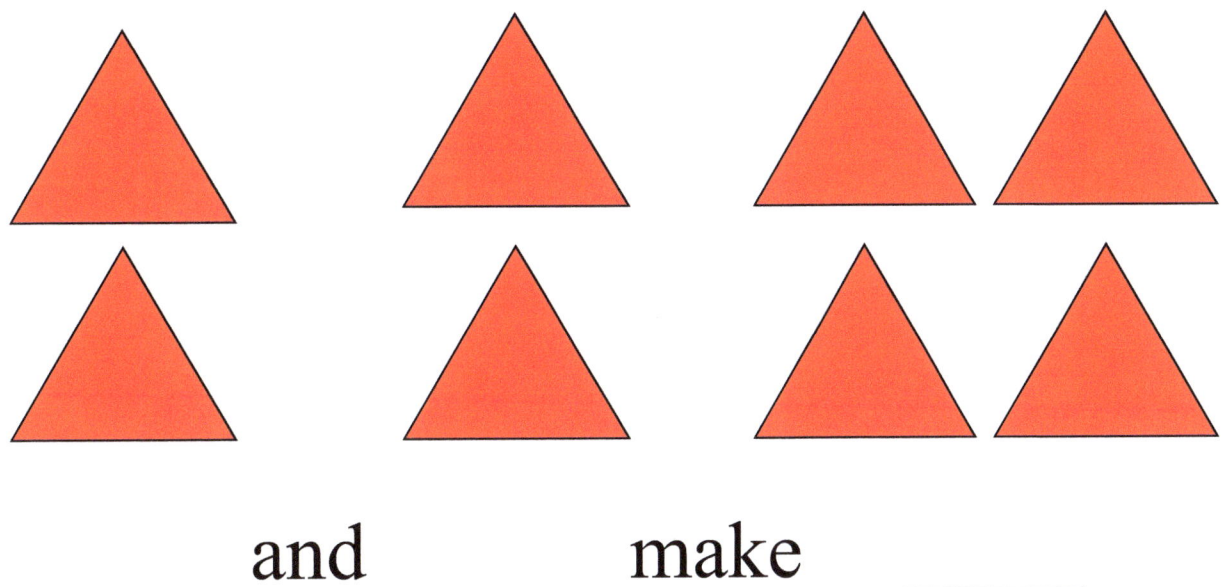

___ and ___ make ____

Numerals 1 to 4

Draw lines to match the numerals to the correct number of shape or shapes.

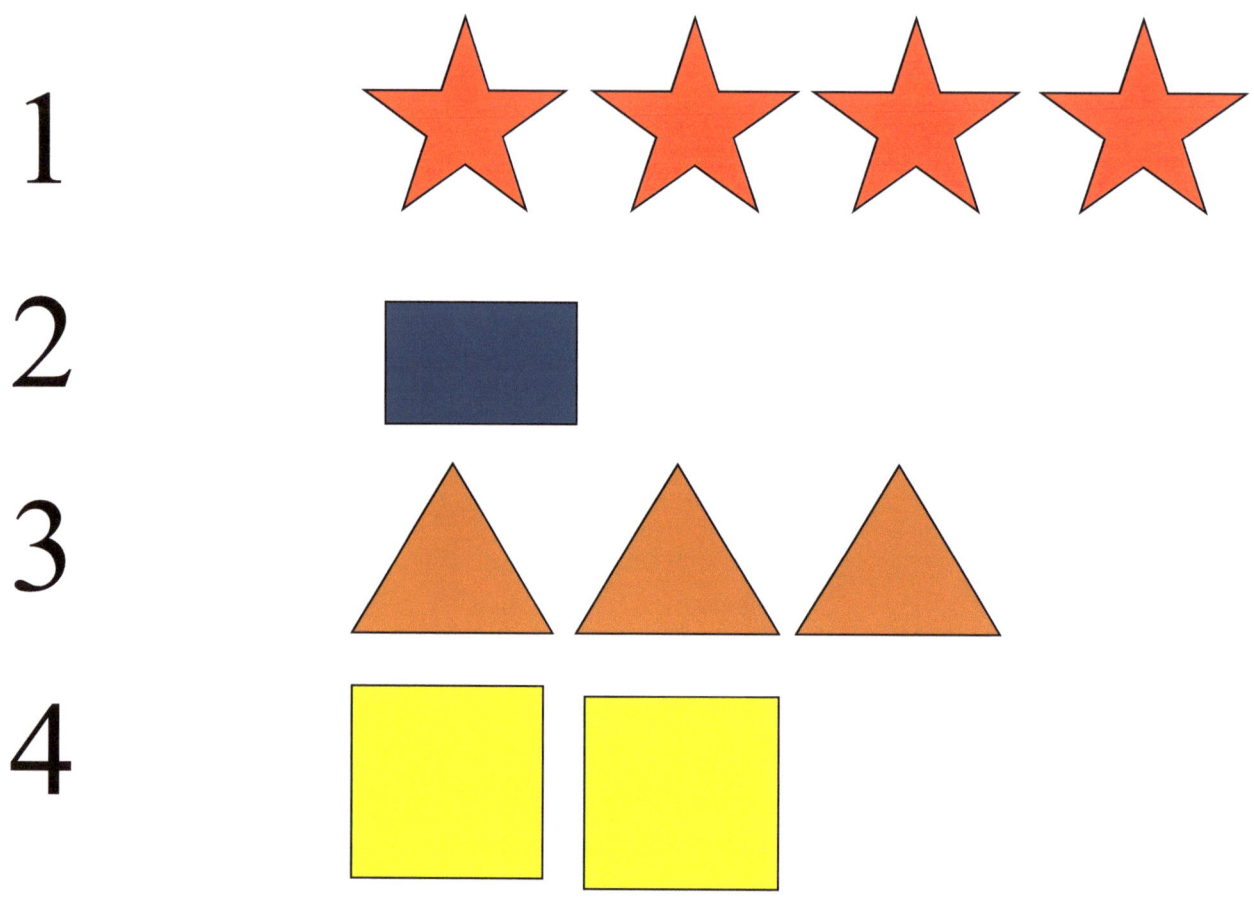

Numerals 1 to 4

Draw a shape or shapes on the lines to match the numerals below.

1 _____

2 _____

3 _____

4 _____

Numerals 1 to 4

1. Colour one rectangle yellow.

2. Colour two triangles blue.

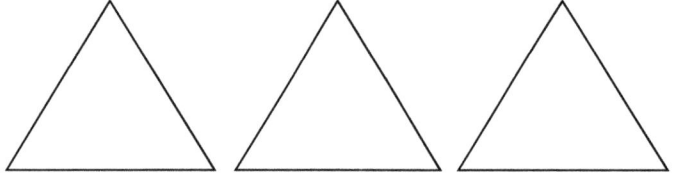

3. Colour three squares red.

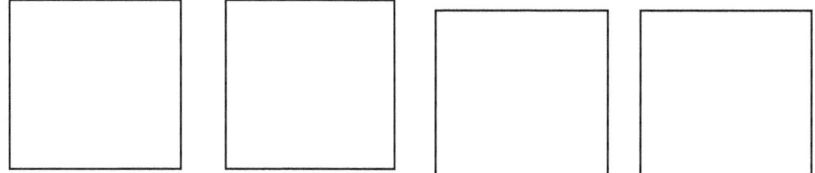

4. Colour four stars orange.

Count the shapes then circle the numeral that tells how many.

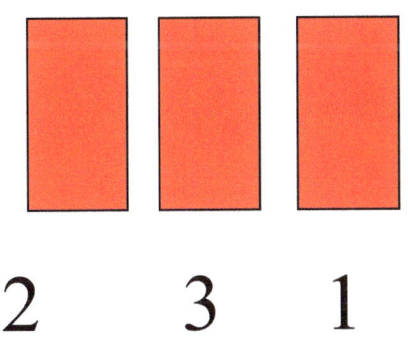

2   3   1

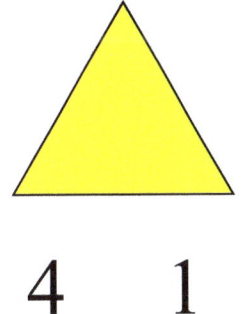

4   1

2   4   1

3   2

Write the numeral which comes next on the lines.

1. 1 ___

2. 1 ___ 3

3. 1 ___ 3 ___

4. 2 ___ 4

Draw and colour four rectangles.

Five purple ovals

Numeral 5

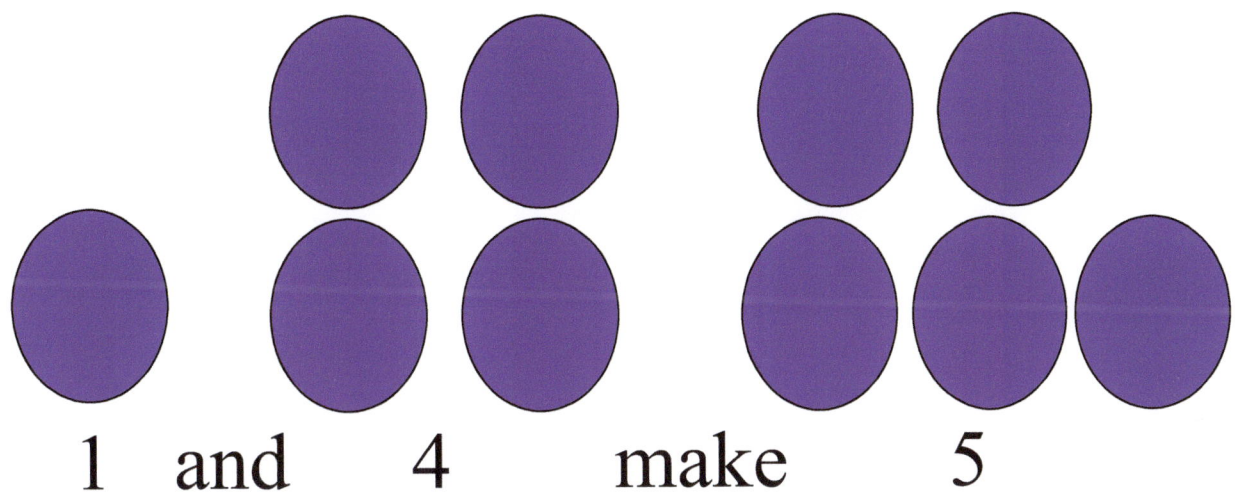

1 and 4 make 5

Draw an oval or ovals on the lines to match the numerals below.

____ and _____ make _____
 1   4   5

Five

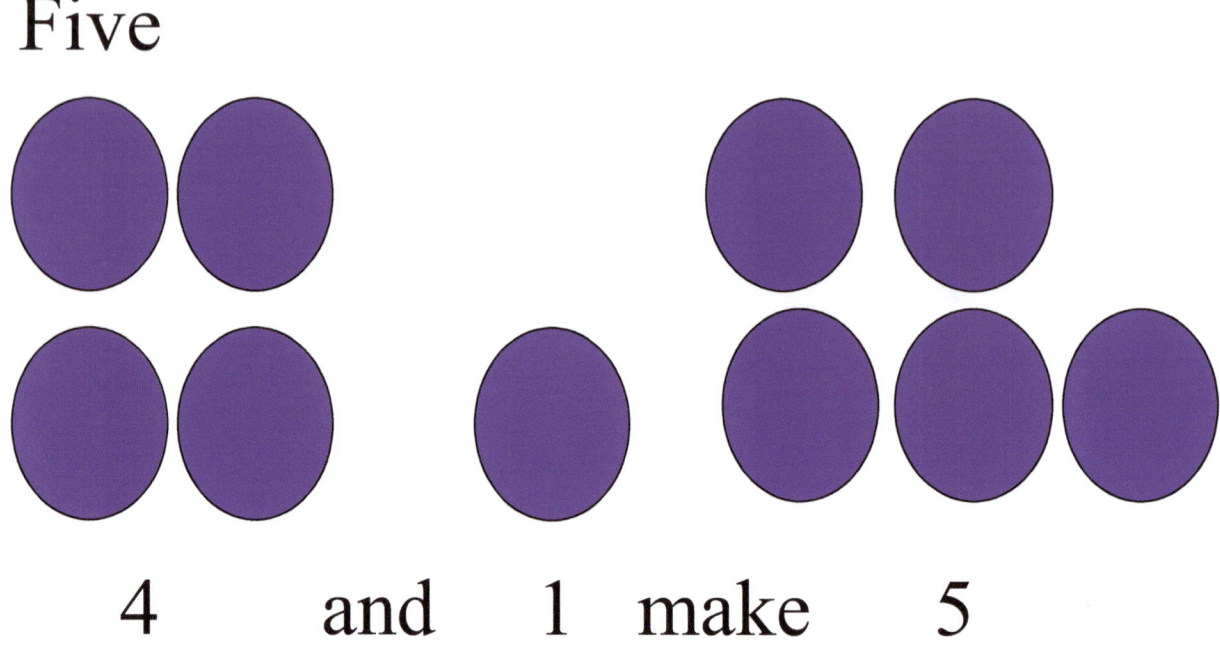

4 and 1 make 5

Draw an oval or ovals on the lines to match the numerals below.

_____ _____ _____
4 and 1 make 5

Numeral 5
Count the ovals then fill in the blanks.

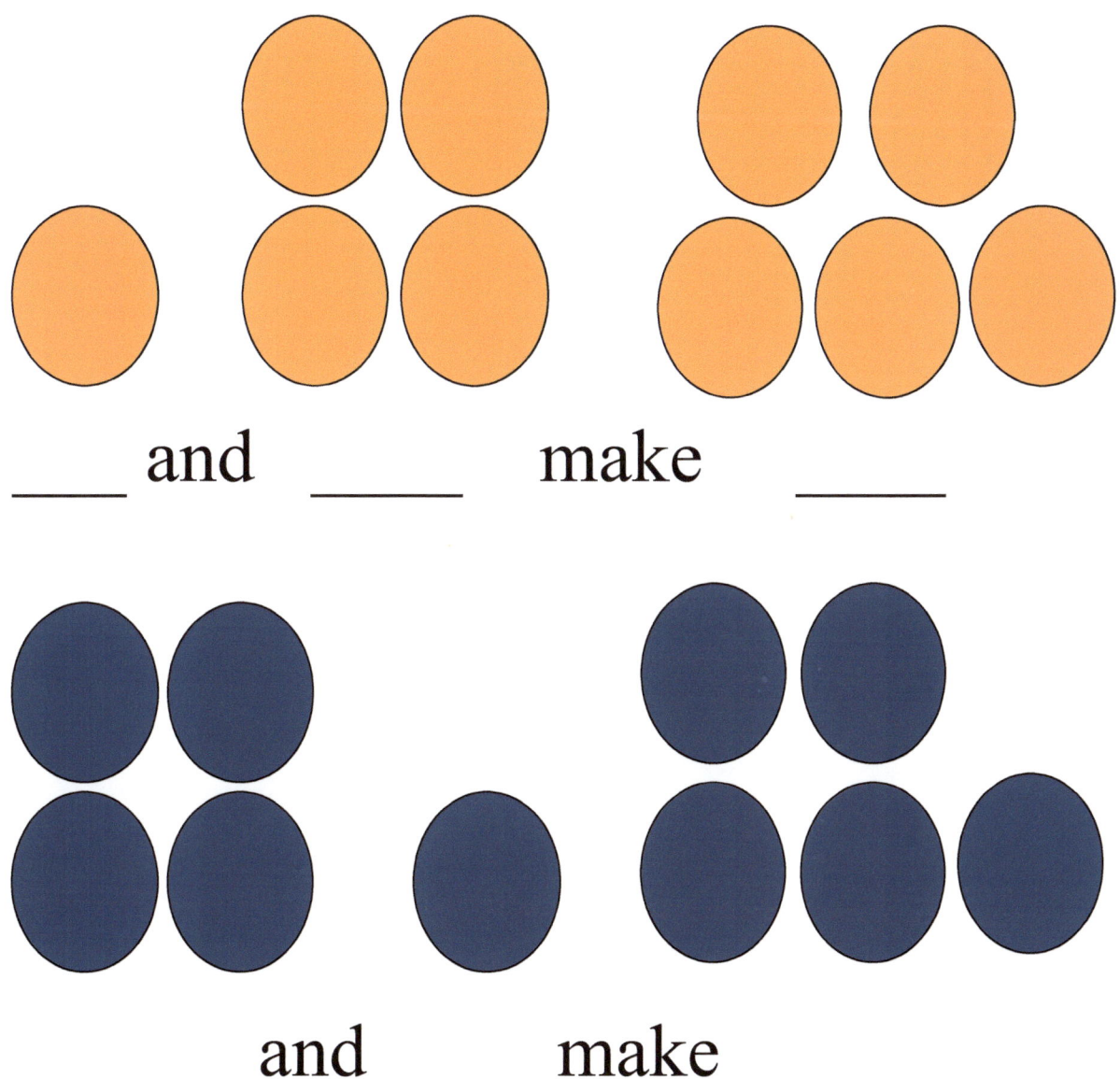

___ and ____ make ____

_____ and ___ make ____

Numeral 5

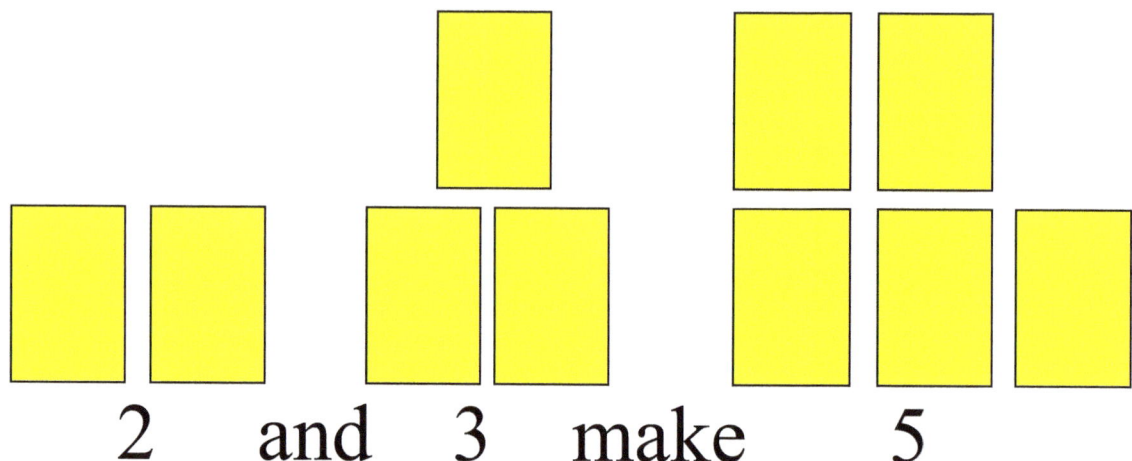

2   and   3   make   5

Draw rectangles on the lines below to match the numerals.

_____   _____   _____

2   and   3   make   5

Five

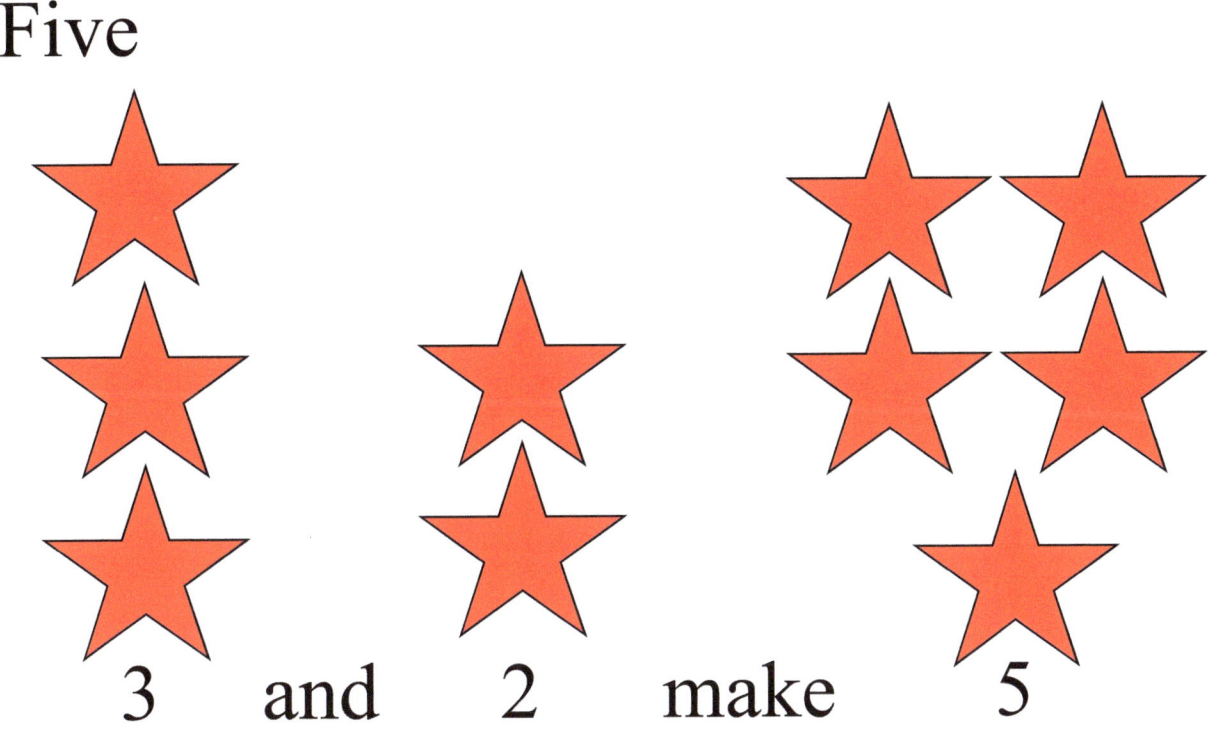

3 and 2 make 5

Draw stars on the lines below to match the numerals.

_____ _____ _____

3 and 2 make 5

Five

Count the ovals then fill in the blanks.

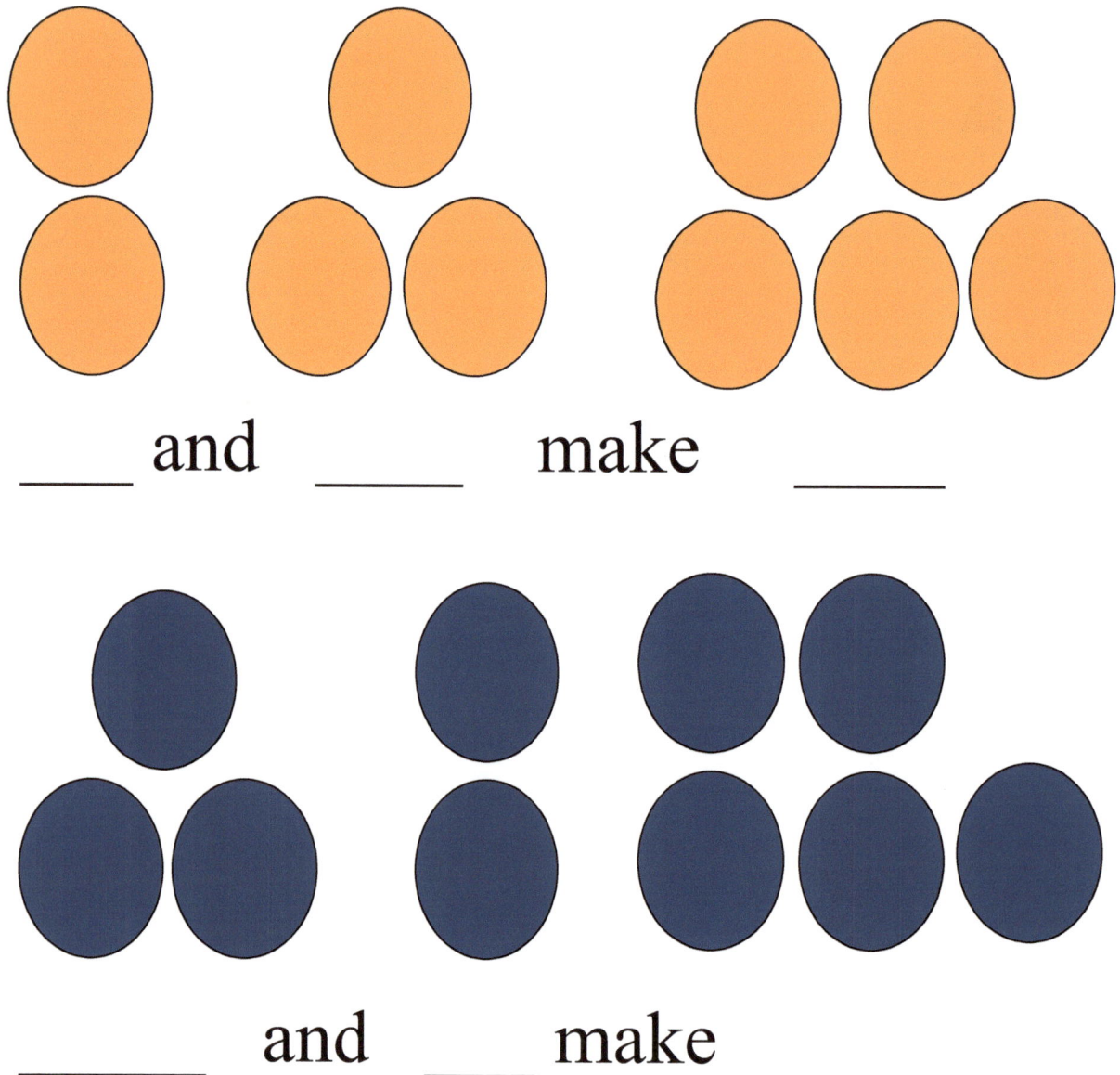

___ and ___ make ___

___ and ___ make ___

Numerals 1 to 5

Draw lines to match the numerals to the correct number of shape or shapes.

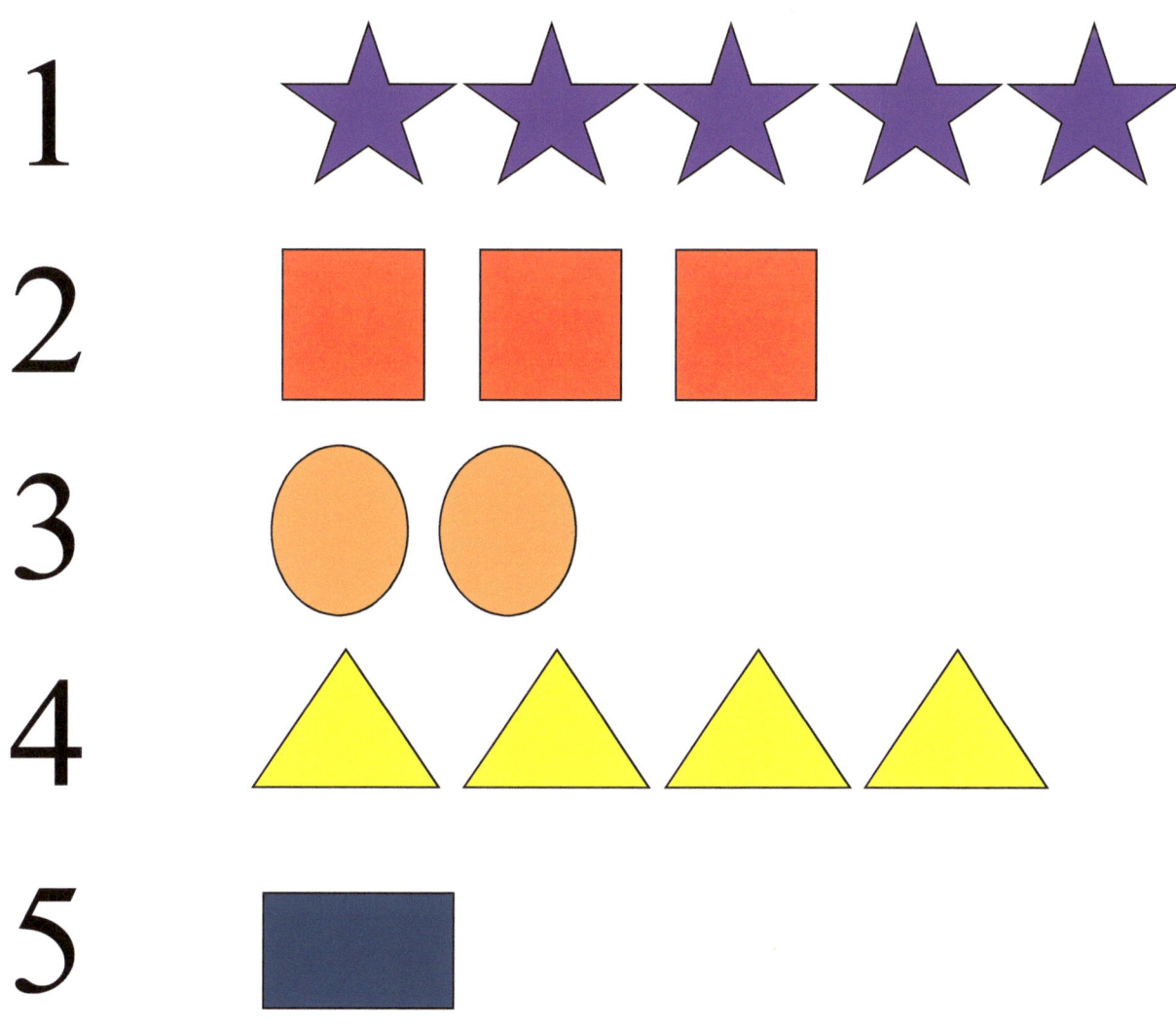

Draw shapes on the lines to match the numerals.

3 _____

4 _____

5 _____

Colouring

1. Colour 2 ovals purple.

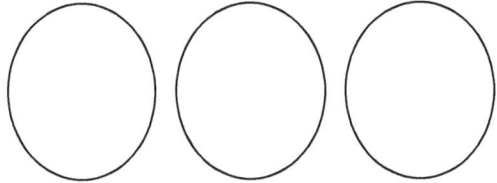

2. Colour 3 triangles blue.

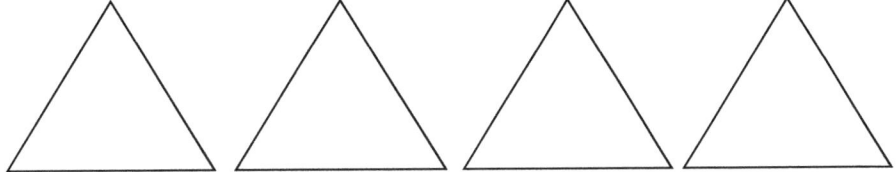

3. Colour 5 squares red.

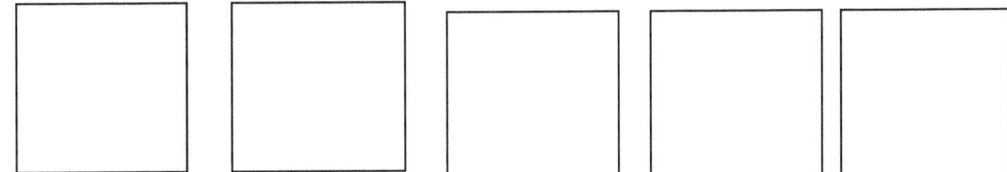

4. Colour 4 stars orange.

Shapes Patterns
Draw and colour the shape that comes next.

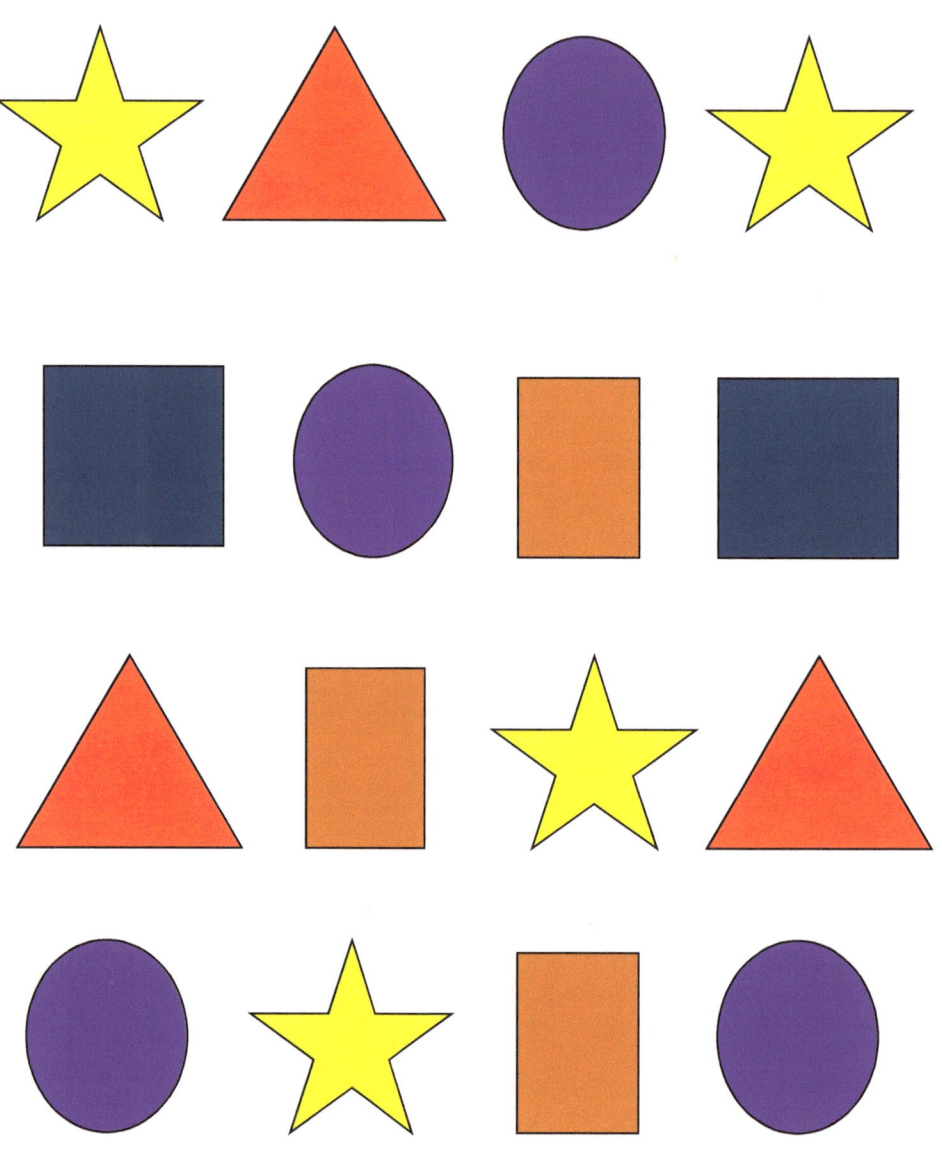

Draw and colour 5 ovals.

Numeral 6

Six green circles

Numeral 6

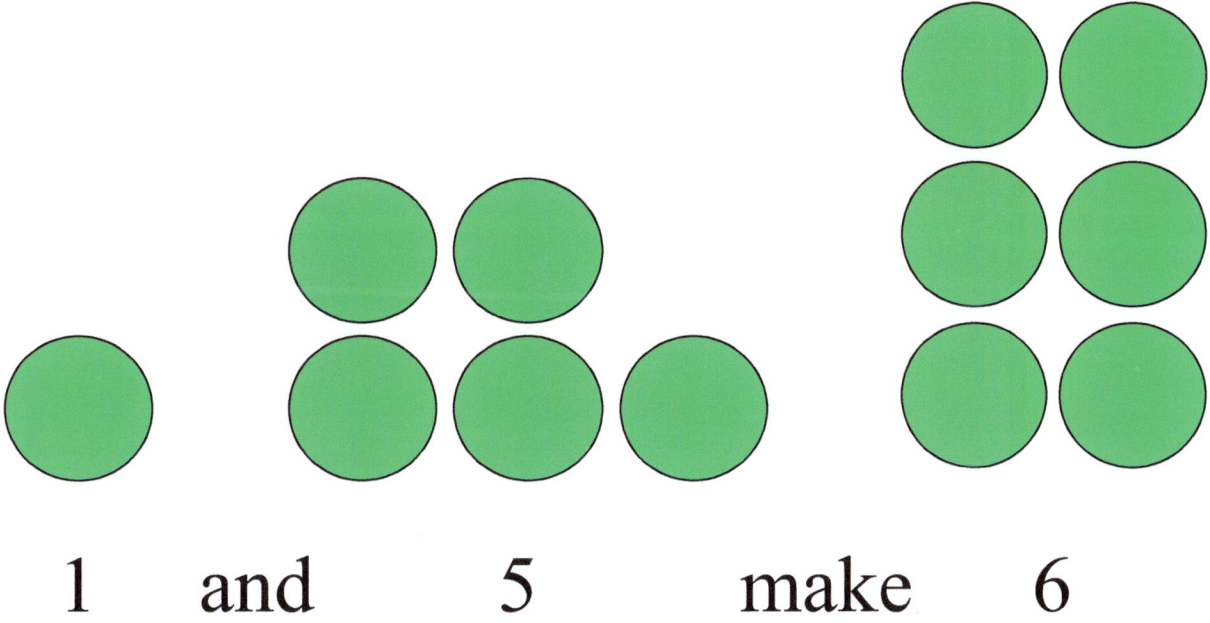

1   and   5   make   6

Draw a shape or shapes on the lines to match the numerals below.

____   _____   _____

1   and   5   make   6

Six

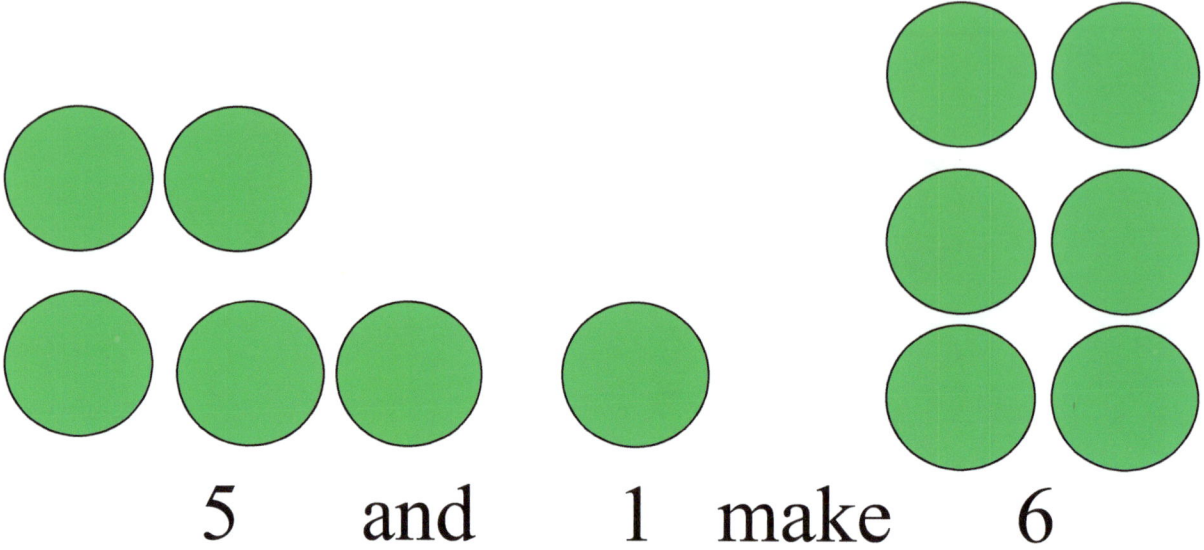

5 and 1 make 6

Draw a circle or circles on the lines to match the numerals below.

_____   ____ _____
      5   and   1   make   6

Count the circles then fill in the blanks below.

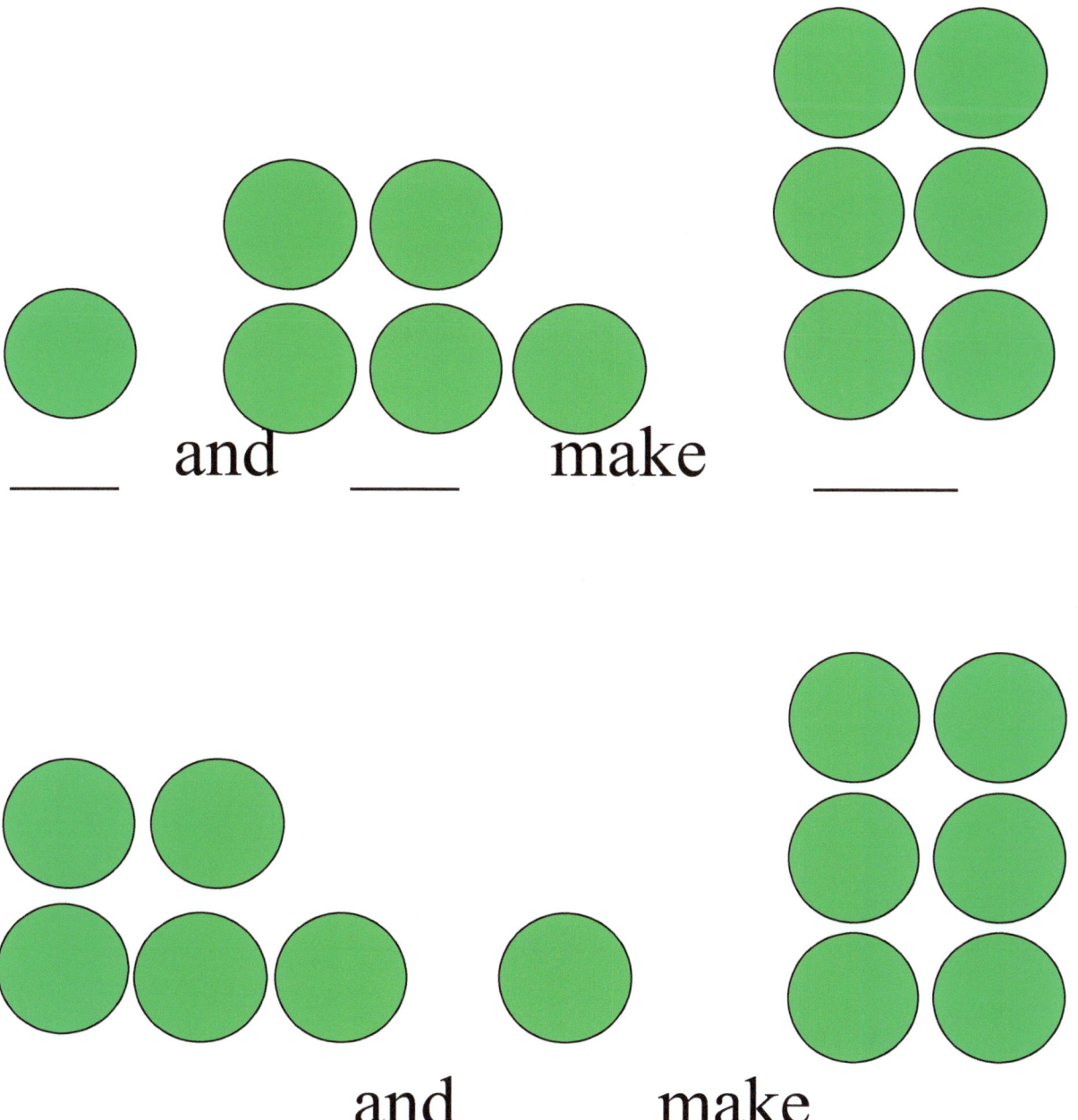

___ and ___ make ____

___ and __ make ___

Numeral 6

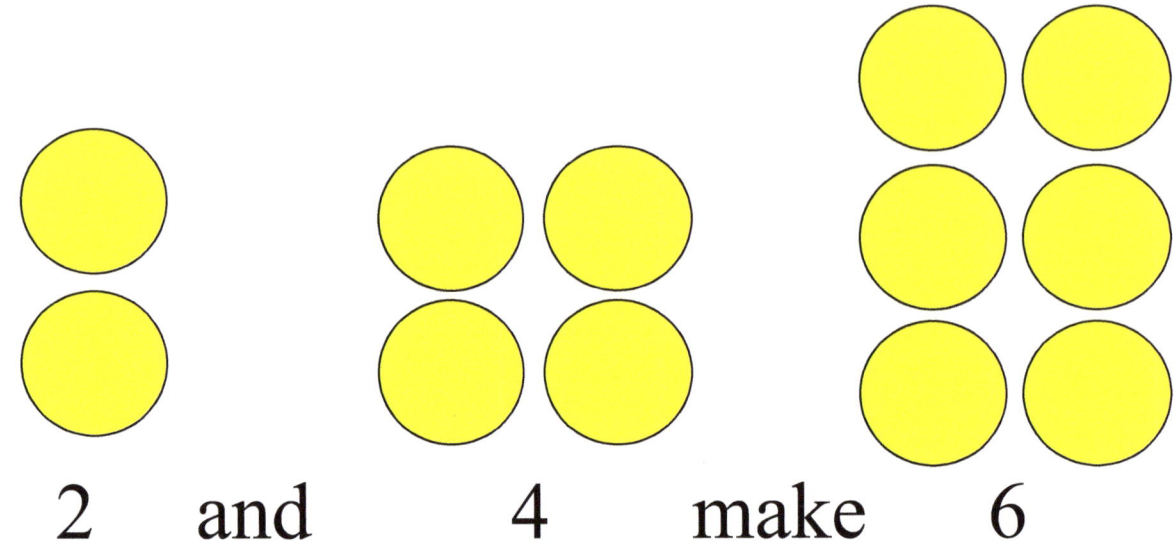

2   and   4   make   6

Draw circles on the lines below to match the numerals.

_____   _____   _____

2   and   4   make   6

Numeral 6

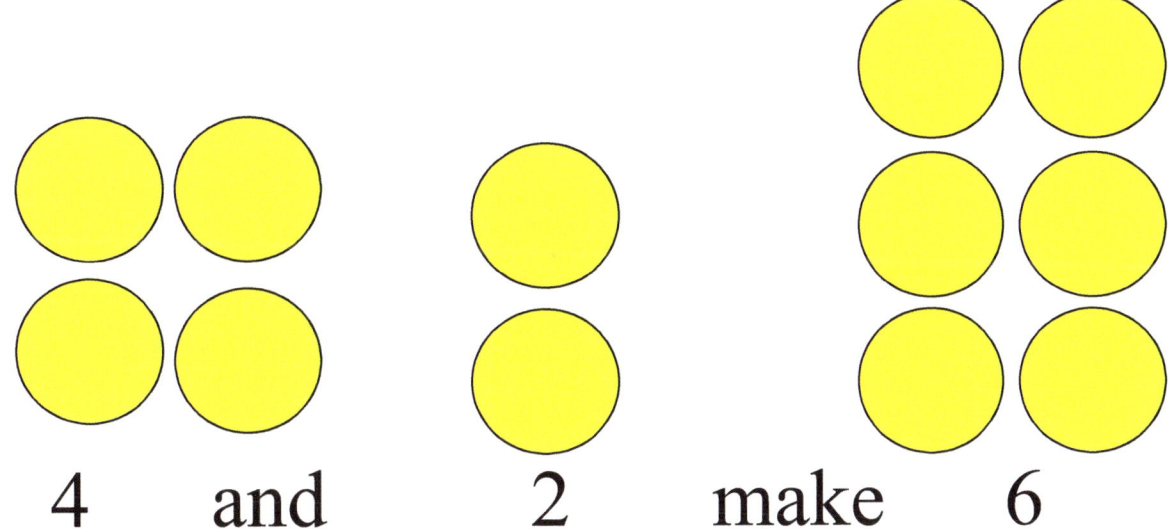

4 and 2 make 6

Draw circles on the lines below to match the numerals.

_____ _____ _____
   4   and   2 make   6

Count the circles then fill in the blanks.

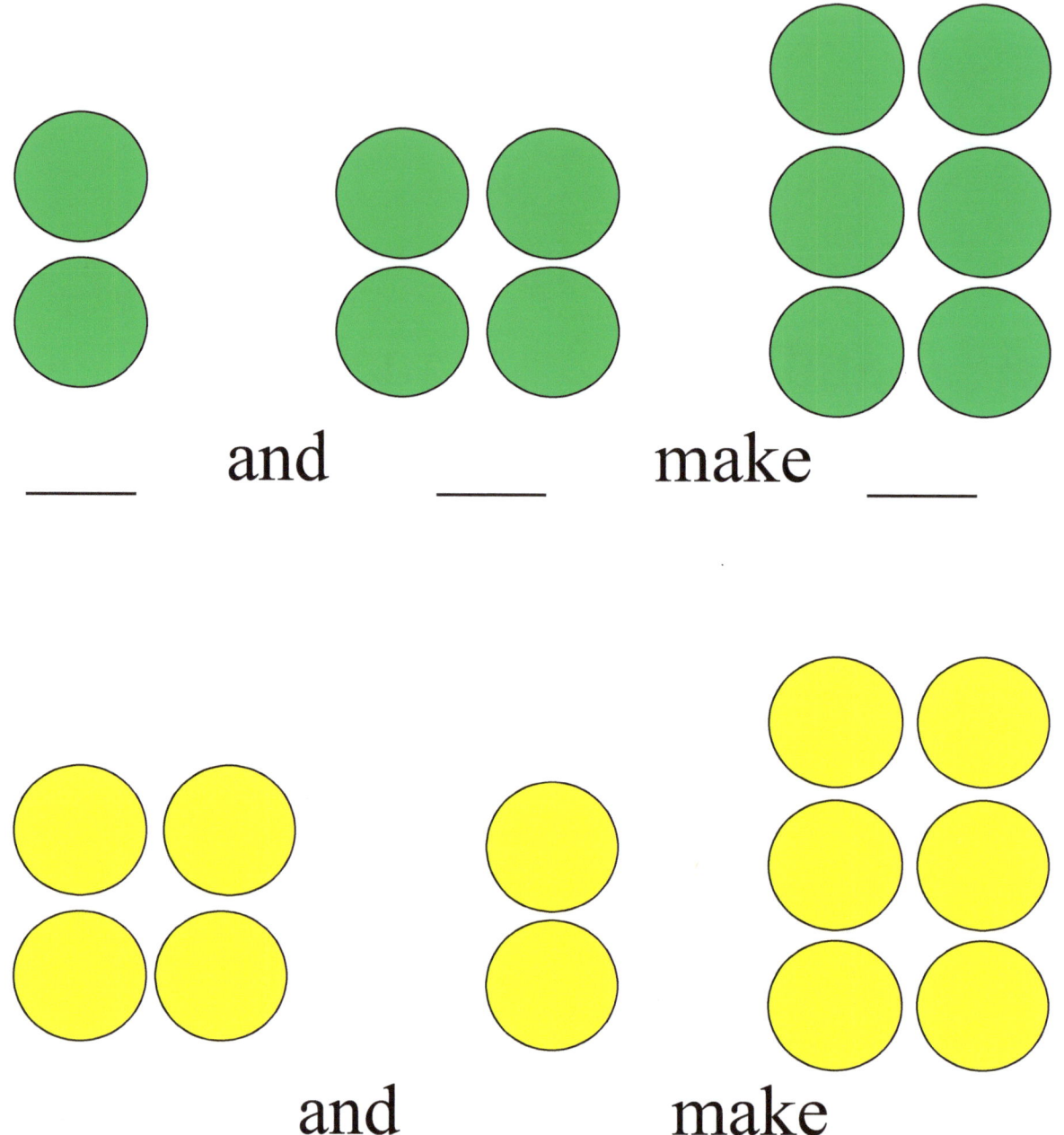

___ and ___ make ___

___ and ___ make ___

Six

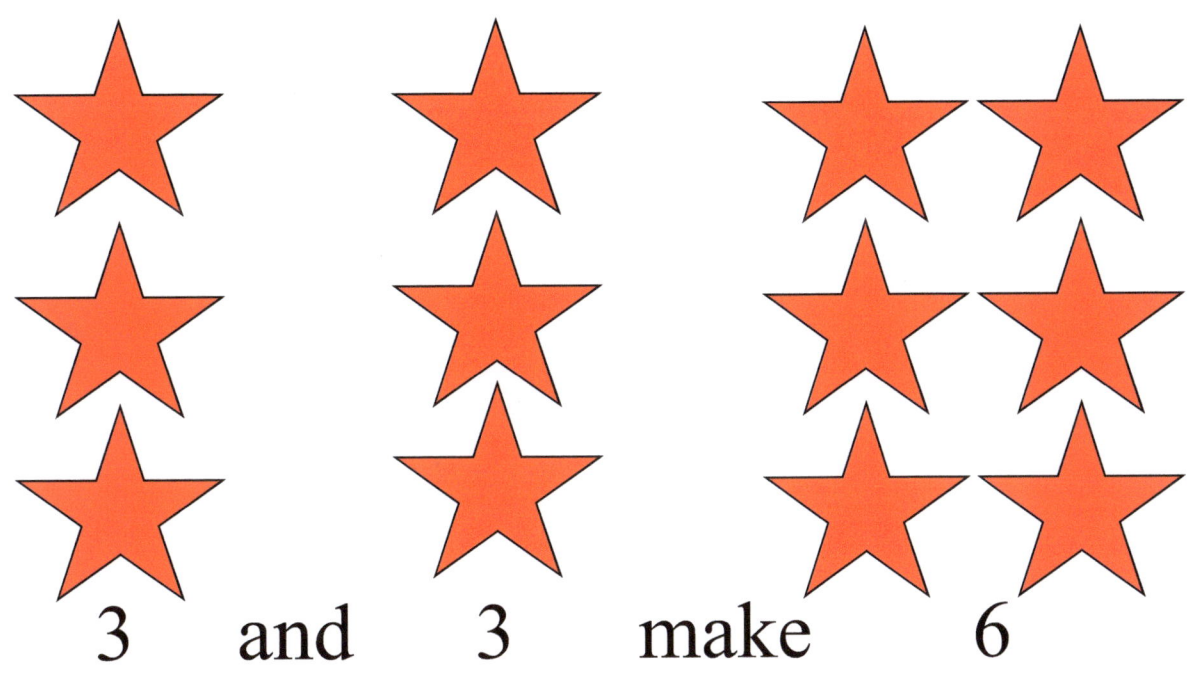

3 and 3 make 6

Draw stars on the lines below to match the numerals.

_____ _____ _____

3 and 3 make 6

Count the stars then fill in the blanks.

___ and ___ make ___

Numerals 1 to 6
Draw lines to match the numerals to the correct number of shape or shapes.

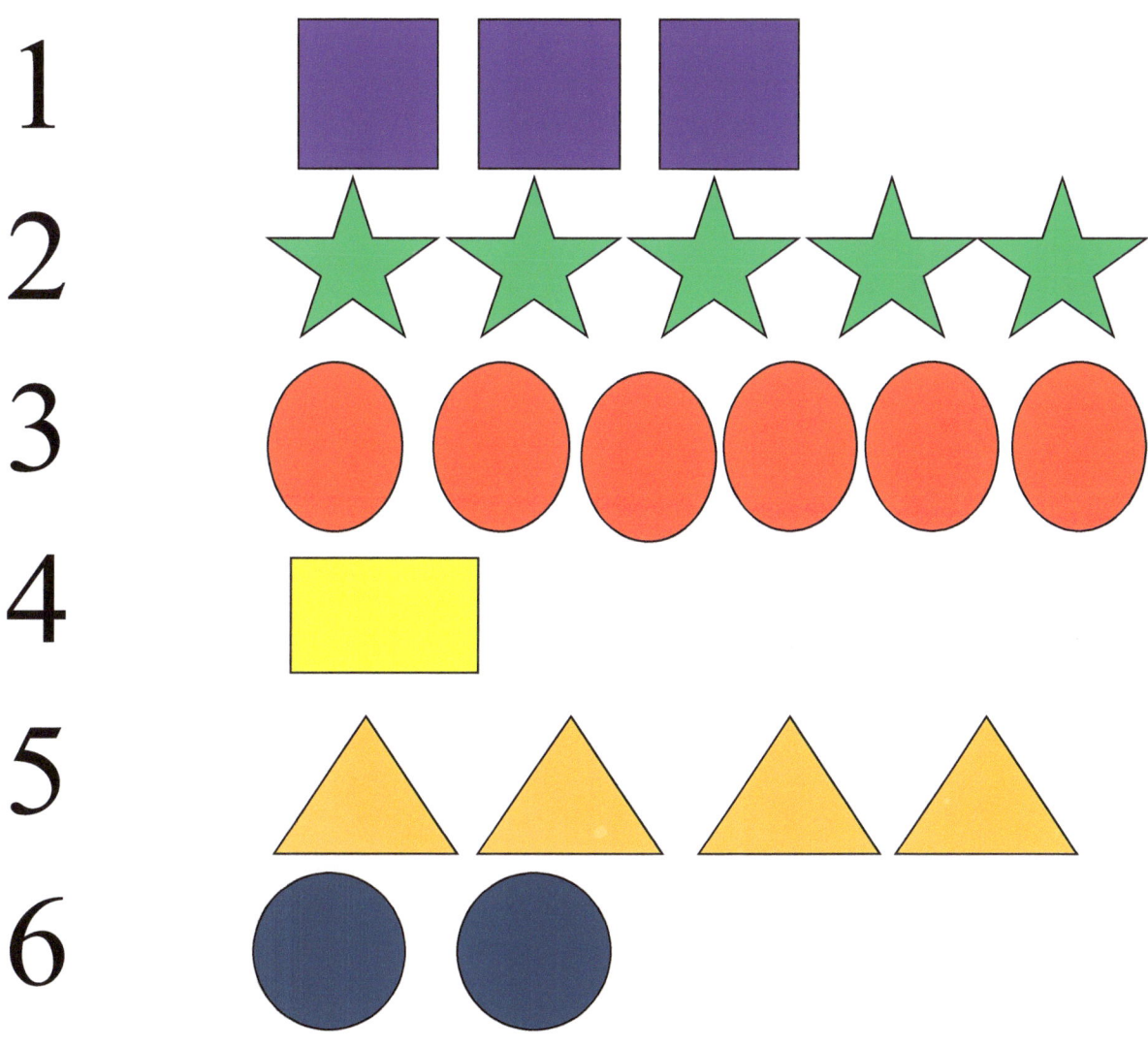

Draw shapes on the lines to match the numerals.

4 _____

5 _____

6 _____

Colouring

1. Colour 2 circles orange.

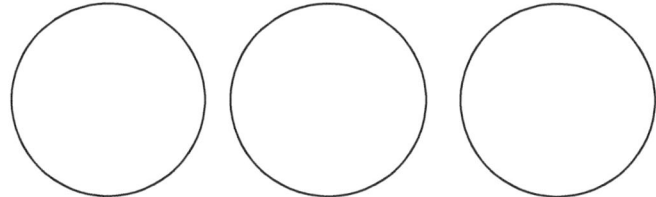

2. Colour 3 squares purple.

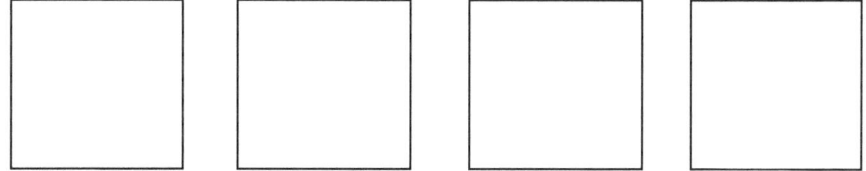

3. Colour 6 ovals green.

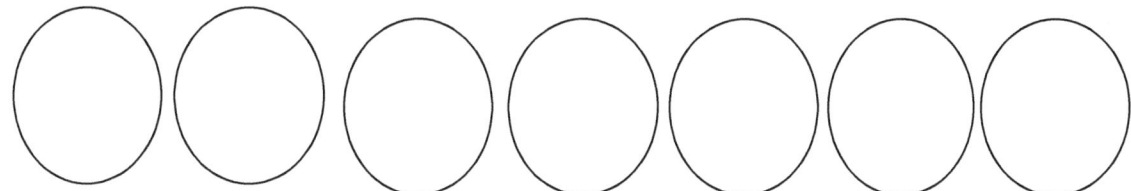

4. Colour 3 stars red.

Shapes Patterns
Draw and colour the shape that comes next.

Match the numerals which are the same.

1	5
2	3
3	1
4	6
5	4
6	2

Match the shapes which are the same.

Draw and colour 6 circles.

Numeral 7

Seven red hearts

Numeral 7

1   and   6   make   7

Draw a heart or hearts on the lines below to match the numerals.

___   _____   _____

1   and   6   make   7

Numeral 7

6   and   1   make   7

Draw a heart or hearts below to match the numerals.

6   and   1   make   7

Count the hearts then fill in the blanks.

___ and ___ make ____

Count the hearts then fill in the blanks.

____ and ____ make ____

Numeral 7

2 and 5 make 7

Draw hearts below to match the numerals.

_____   _____   _____

2 and 5 make 7

Numeral 7

5 and 2 make 7

Draw hearts below to match the numerals.

_____ ____ _____

5 and 2 make 7

Count the circles then fill in the blanks below.

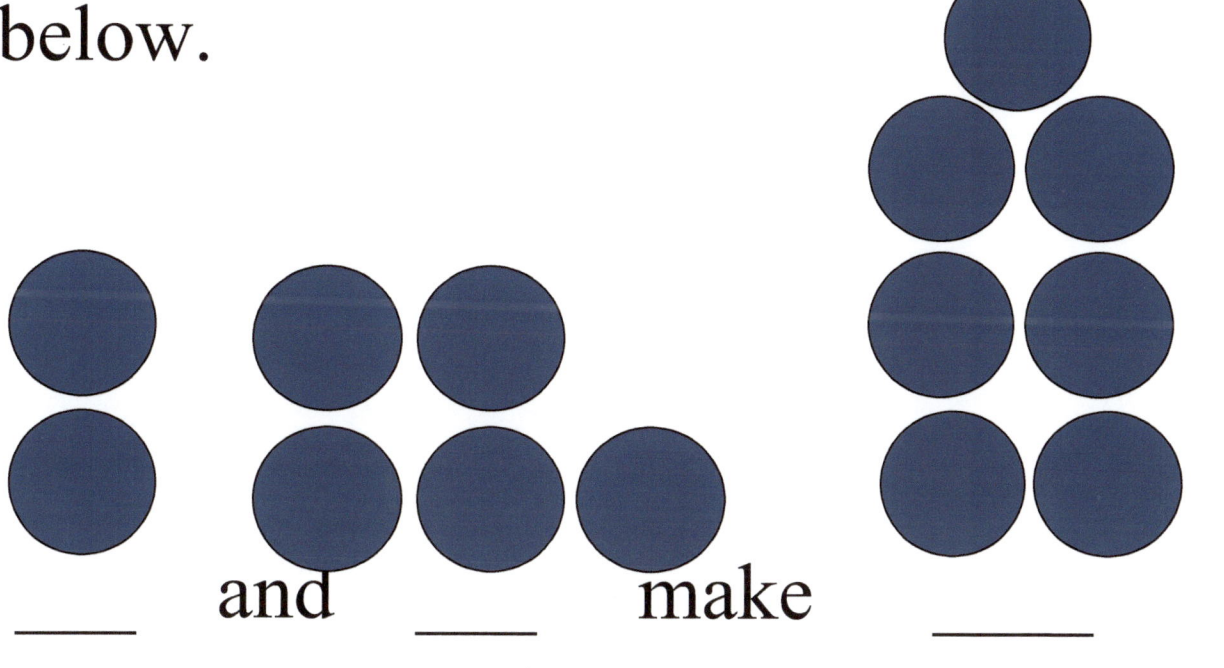

___ and ___ make ___

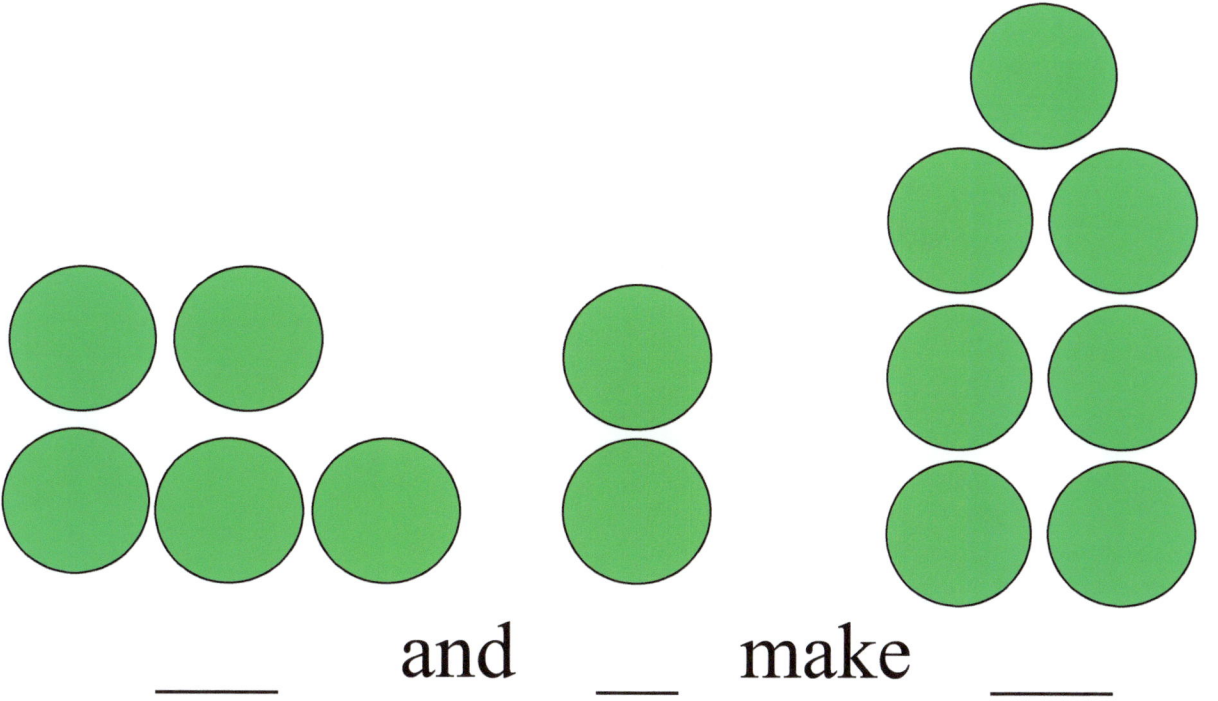

___ and ___ make ___

Numeral 7

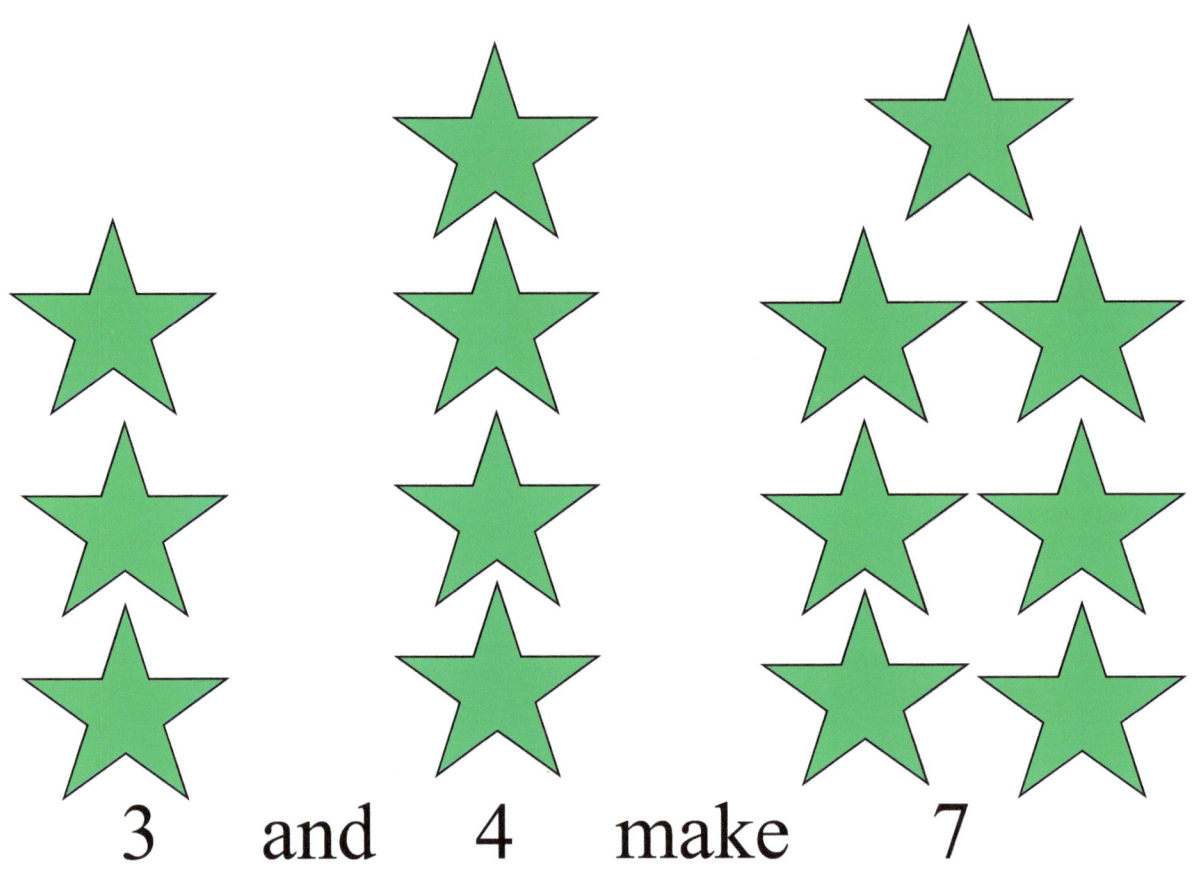

3 and 4 make 7

Numeral 7

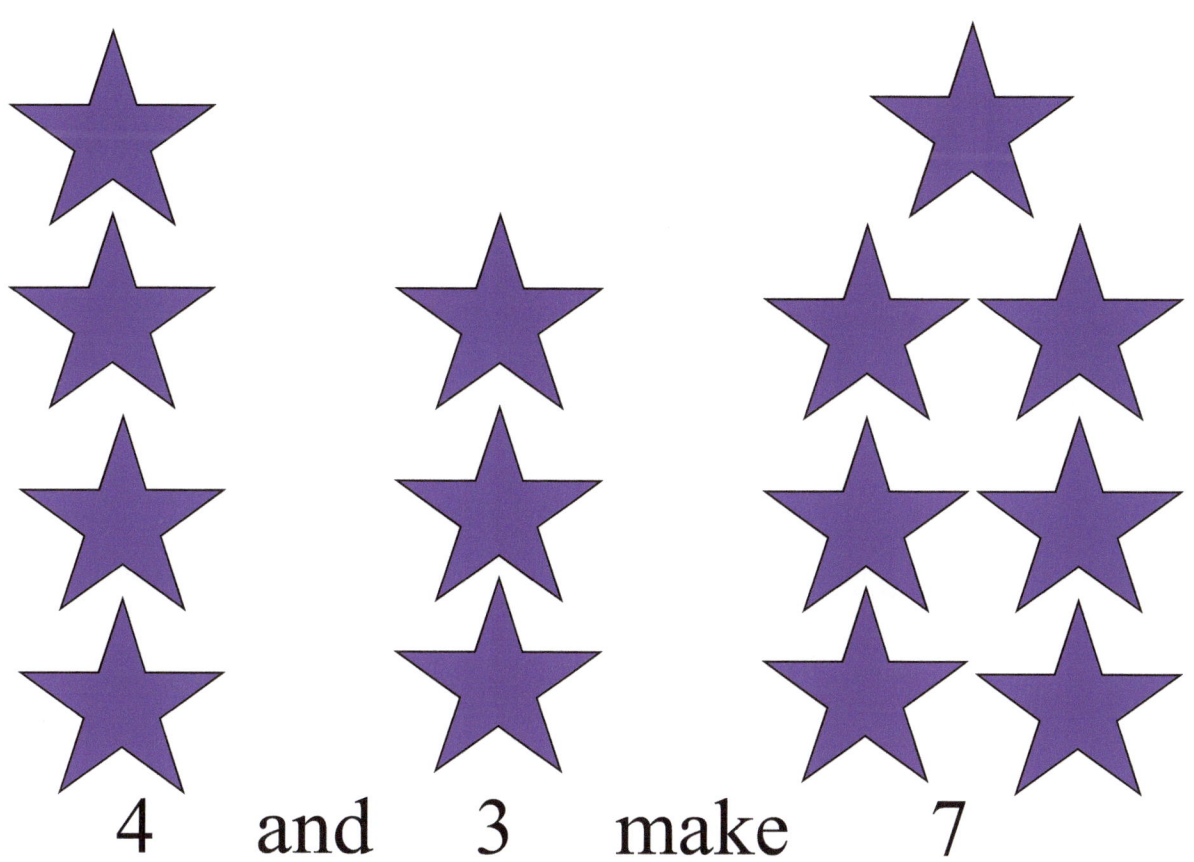

4 and 3 make 7

Count the triangles then fill in the blanks below.

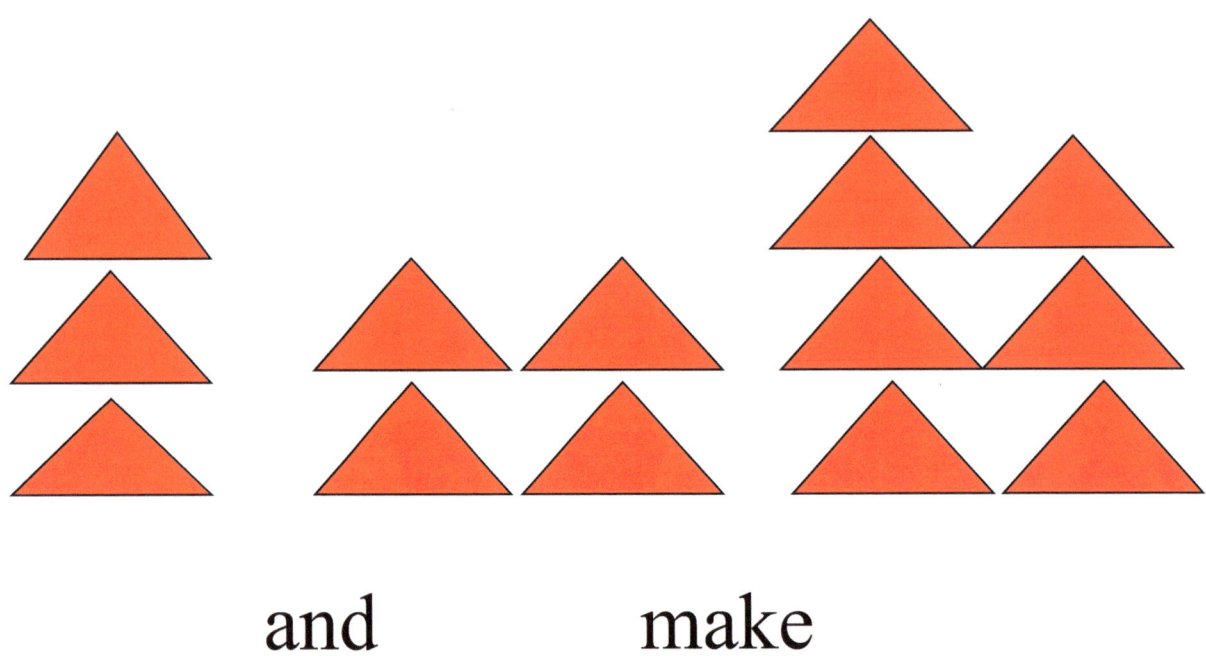

____ and ____ make ____

____ and ____ make ____

Draw lines to match the numerals to the correct number of shape or shapes.

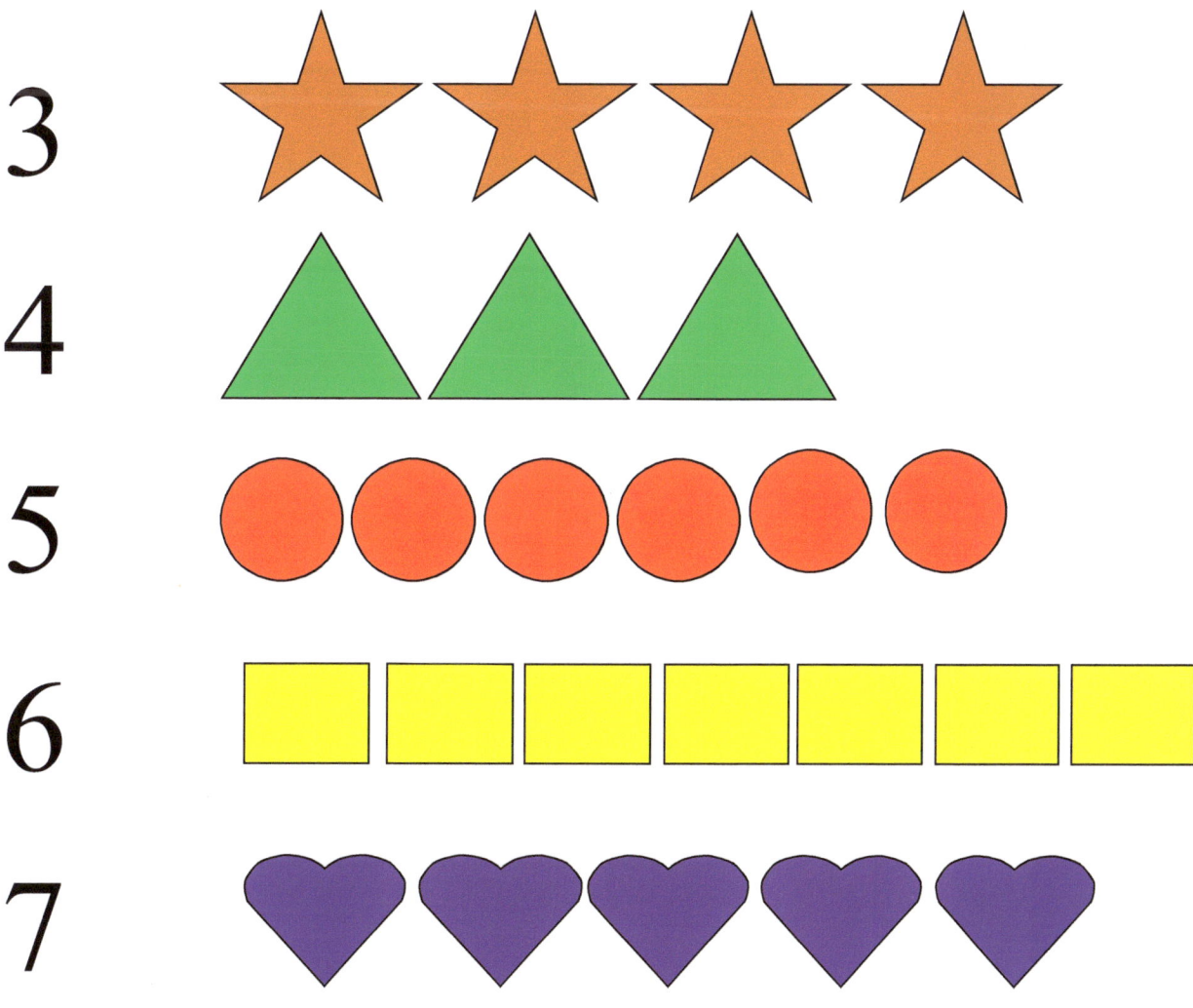

Draw shapes on the lines to match the following numerals.

2 _____

3 _____

4 _____

5 _____

6 _____

7 _____

Colouring

1. Colour one triangle purple.

2. Colour two rectangles yellow.

3. Colour three stars red.

4. Colour four squares orange.

5. Colour five circles blue.

6. Colour six hearts green.

Write the numeral which comes next on the lines.

1. 1 ___ 3
2. 1 ___ 3 ___
3. 2 ___ 4 ___
4. 3 ___ 5 ___
5. 4 ___ 6 ___

Shapes Patterns

Colour the shape that comes next.

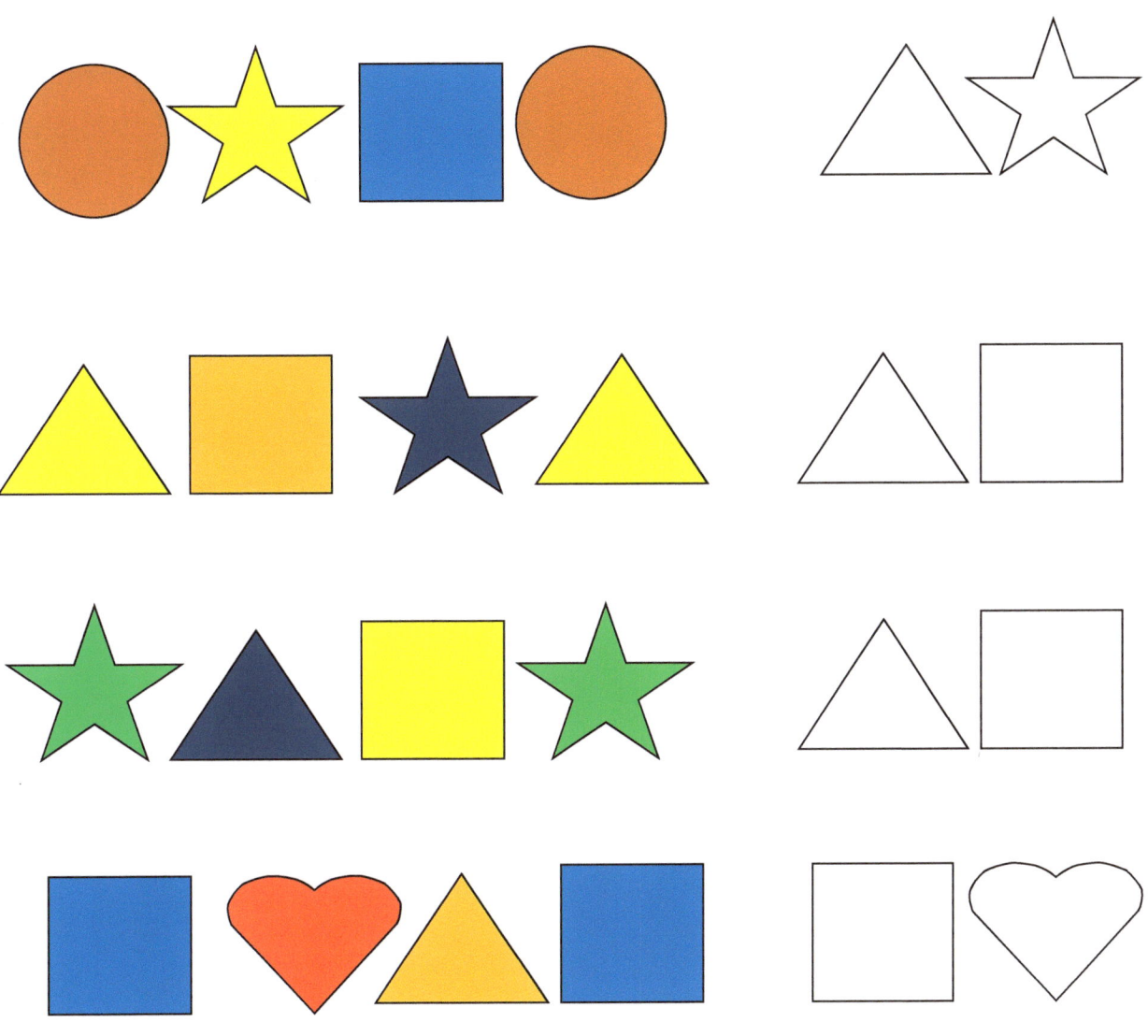

Count the shapes then circle the numeral which tell how many.

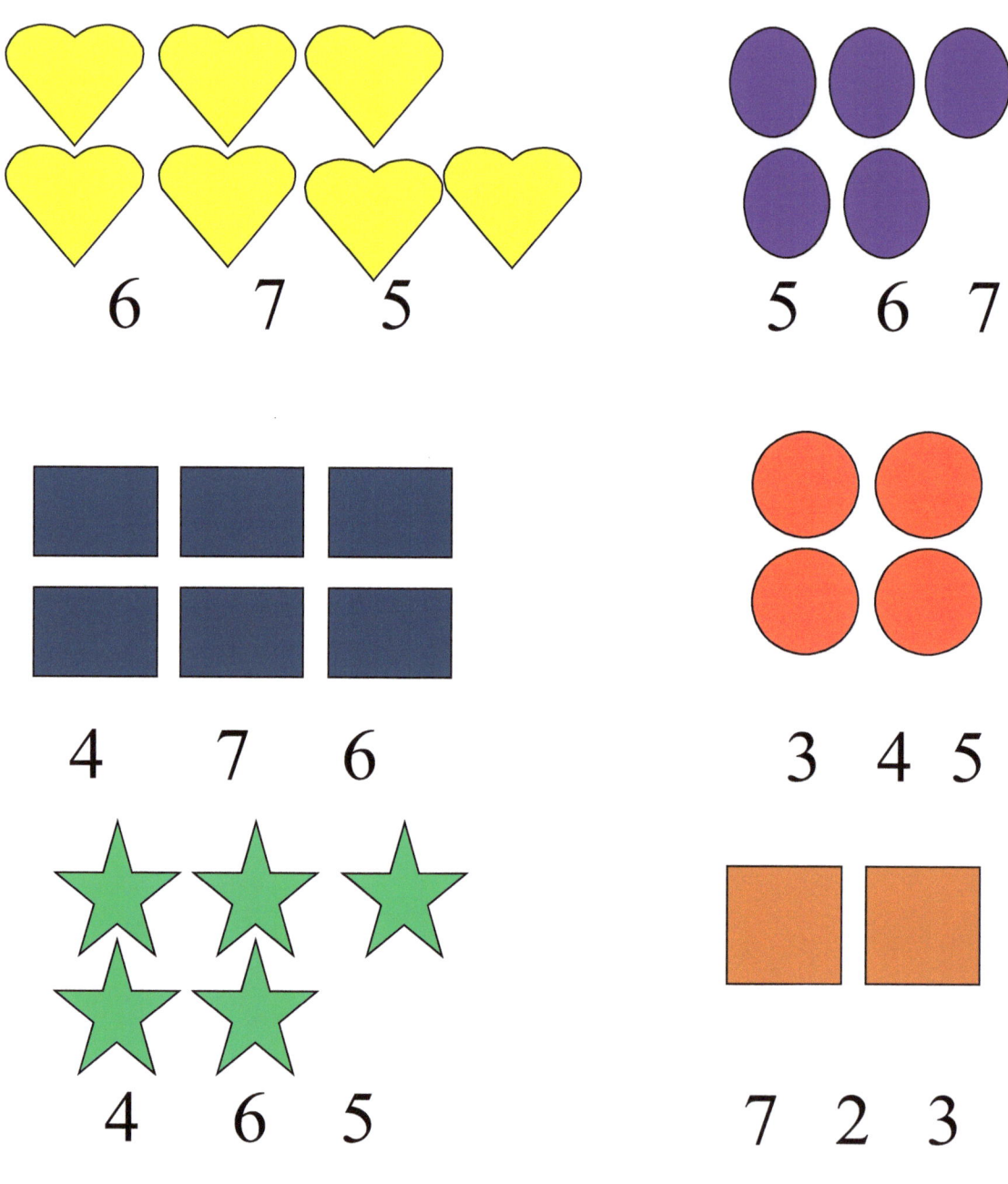

Draw and colour 7 hearts.

Numeral 8

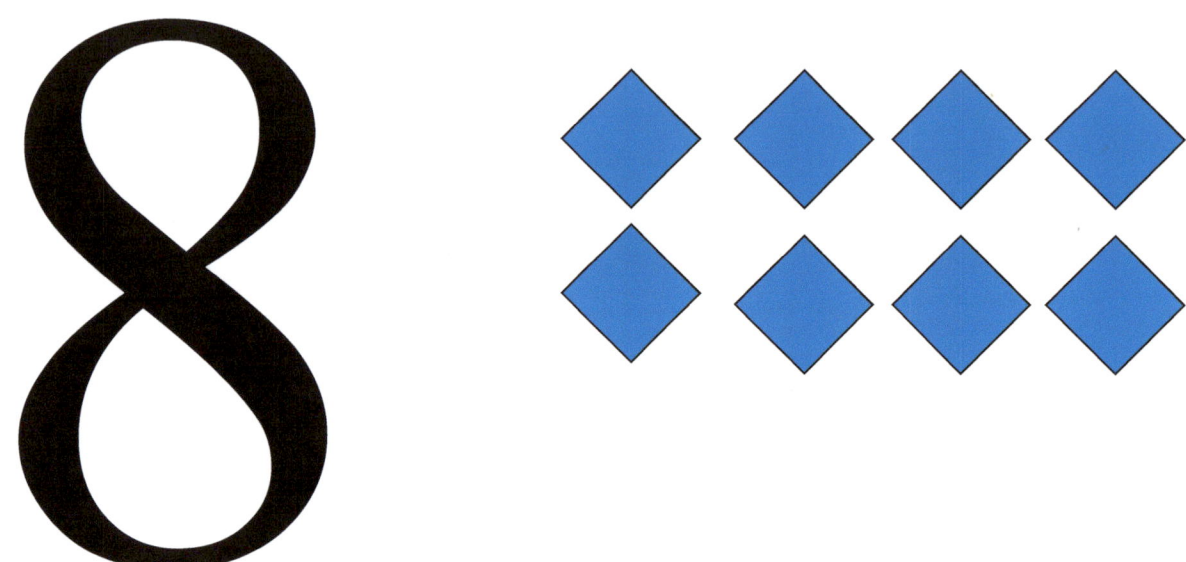

Eight blue diamonds

Numeral 8

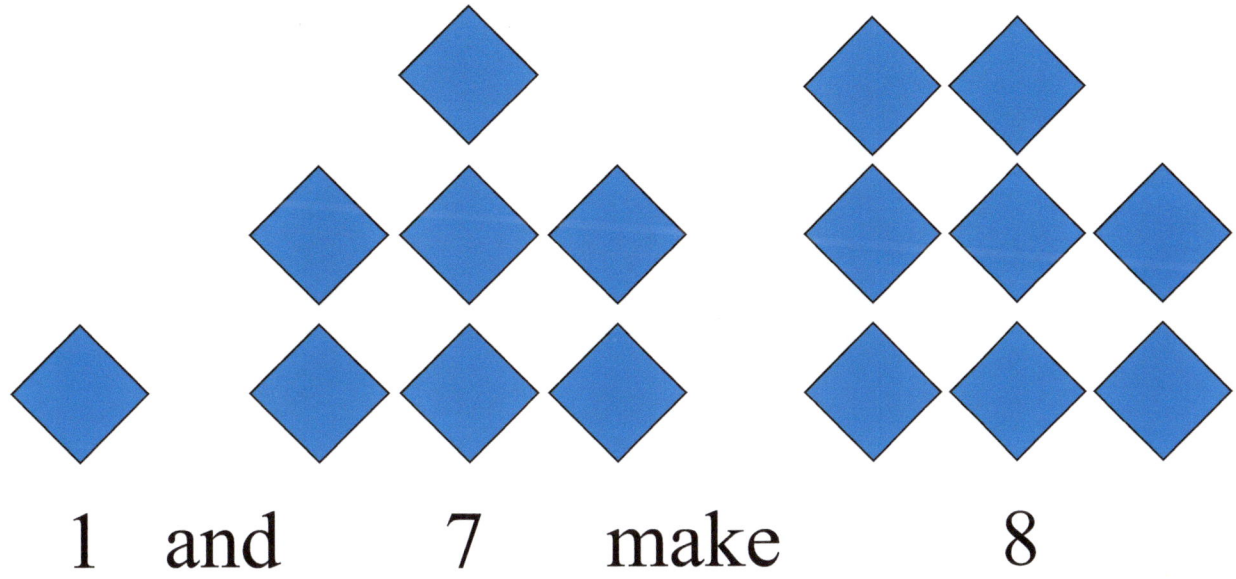

1   and   7   make   8

Draw a diamond or diamonds on the lines to match the numerals below.

_____   _____   _____

1   and   7   make   8

Numeral 8

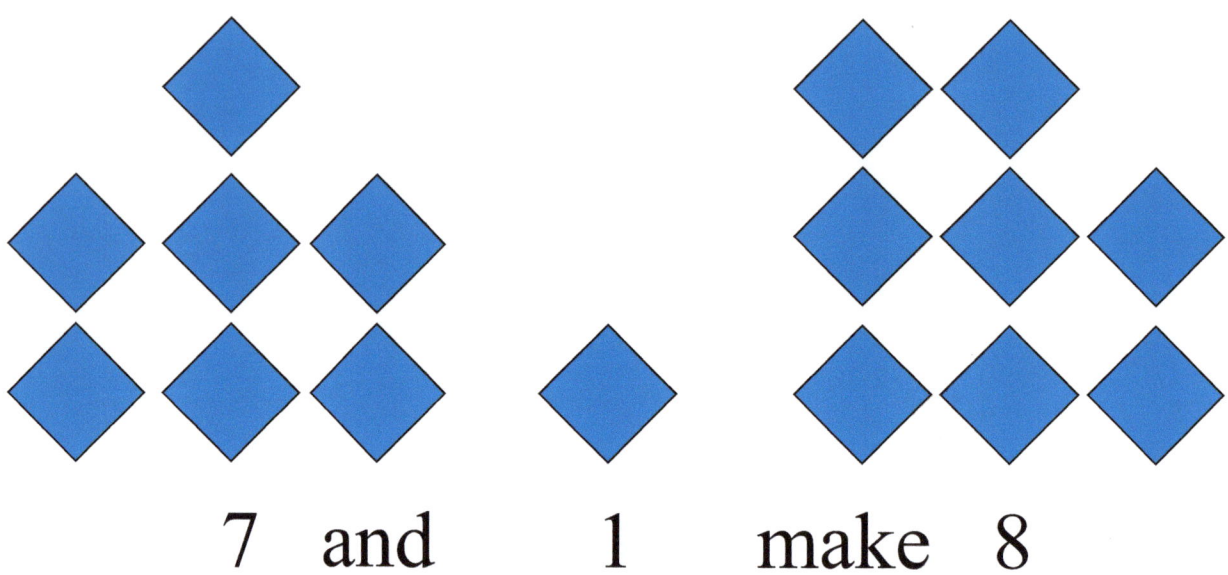

7 and 1 make 8

Draw diamonds on the lines to match the numerals below.

_____   ____ _____

7 and 1 make 8

Numeral 8

Count the diamonds then fill in the blanks.

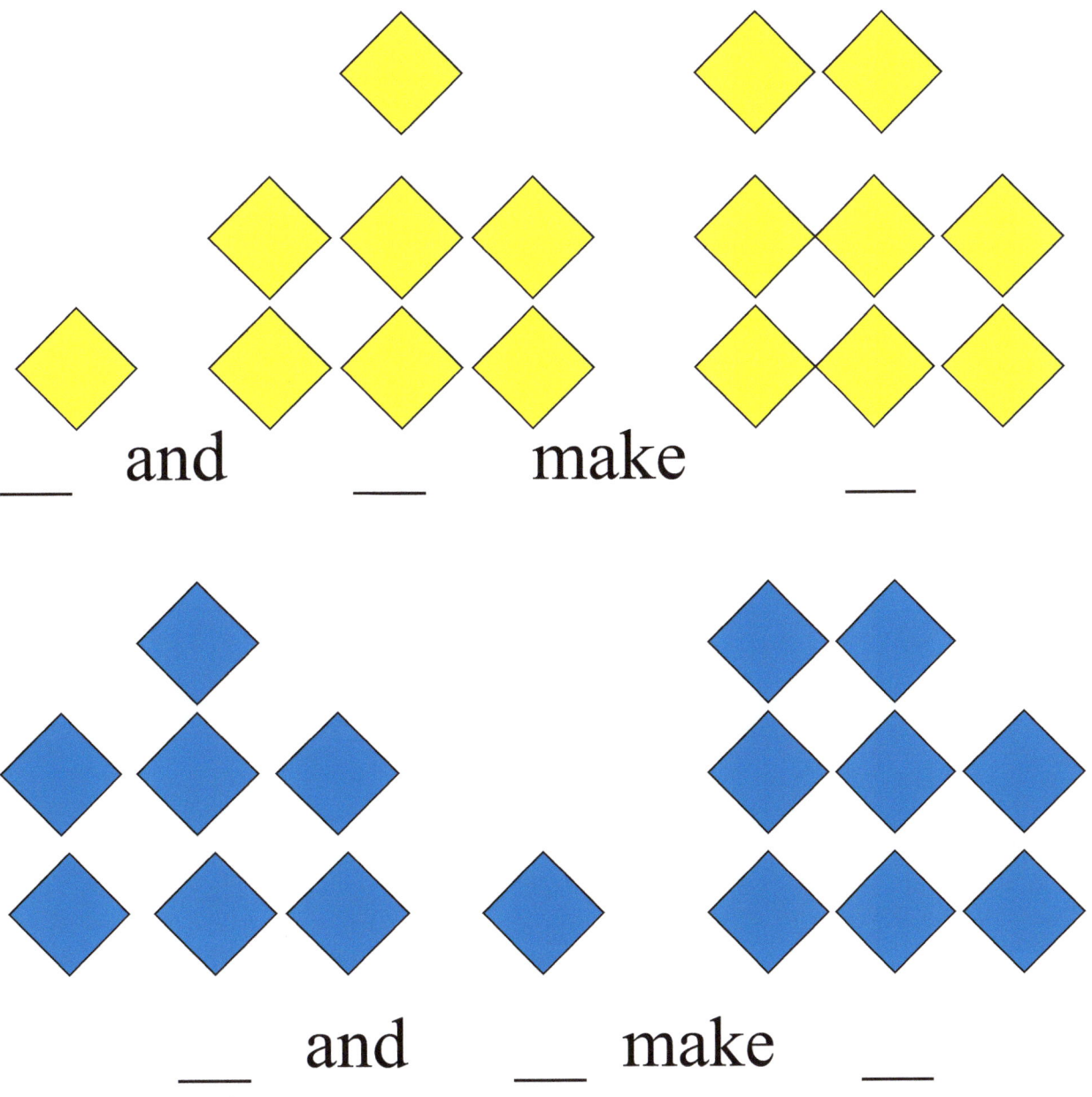

Numeral 8

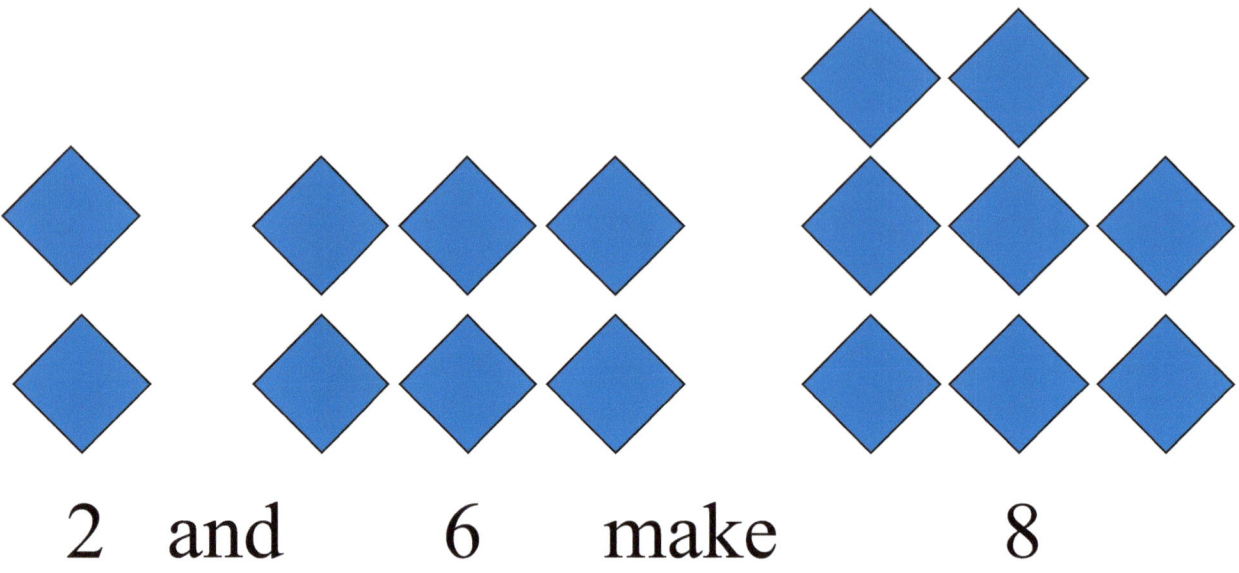

2 and   6   make   8

Draw diamonds on the lines to match the numerals below.

_____   _____   _____

2 and   6   make   8

Numeral 8

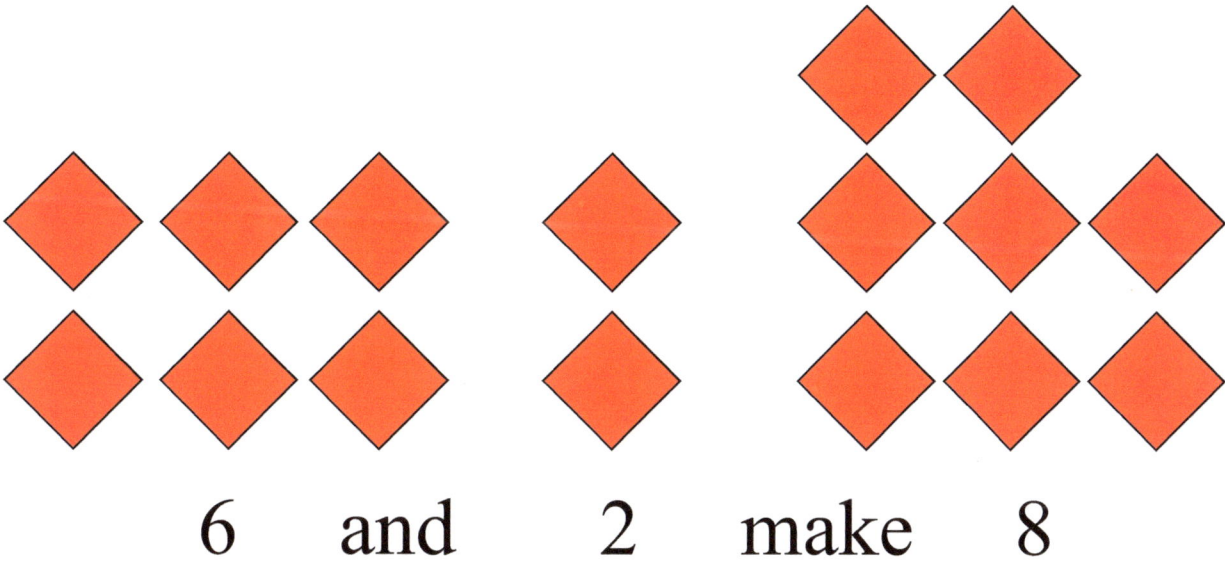

6 and 2 make 8

Draw diamonds on the lines to match the numerals below.

_____   _____   _____

6 and   2 make   8

Numeral 8

Count the diamonds then fill in the blanks.

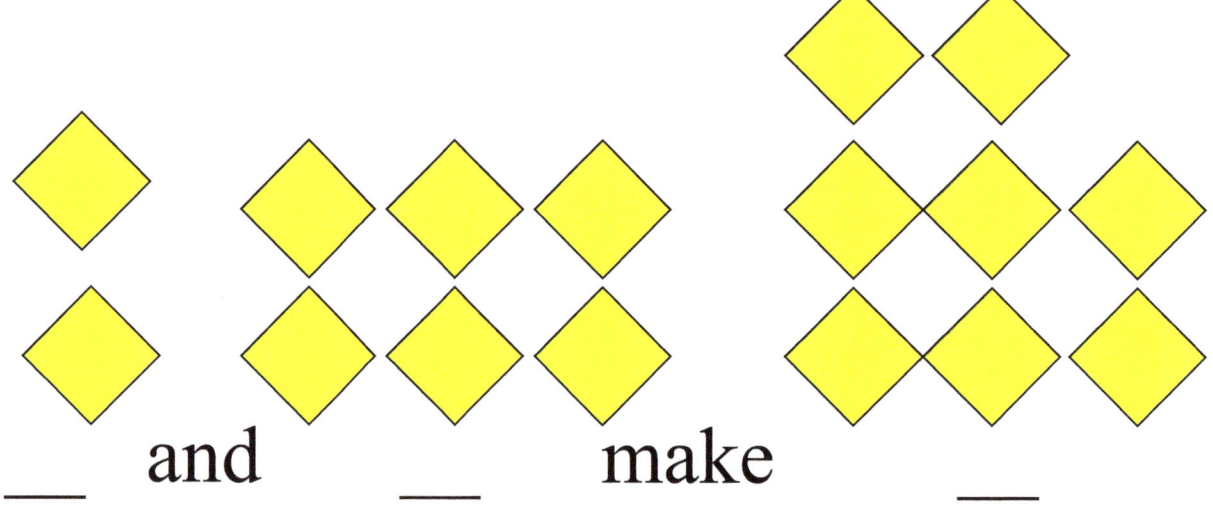

__ and __ make __

__ and __ make __

Numeral 8

3   and   5   make   8

Draw hearts below to match the numerals.

_____   _____   _____

3   and   5   make   8

Numeral 8

5 and 3 make 8

Draw hearts below to match the numerals.

_____ _____ _____

5 and 3 make 8

Count the circles then fill in the blanks below.

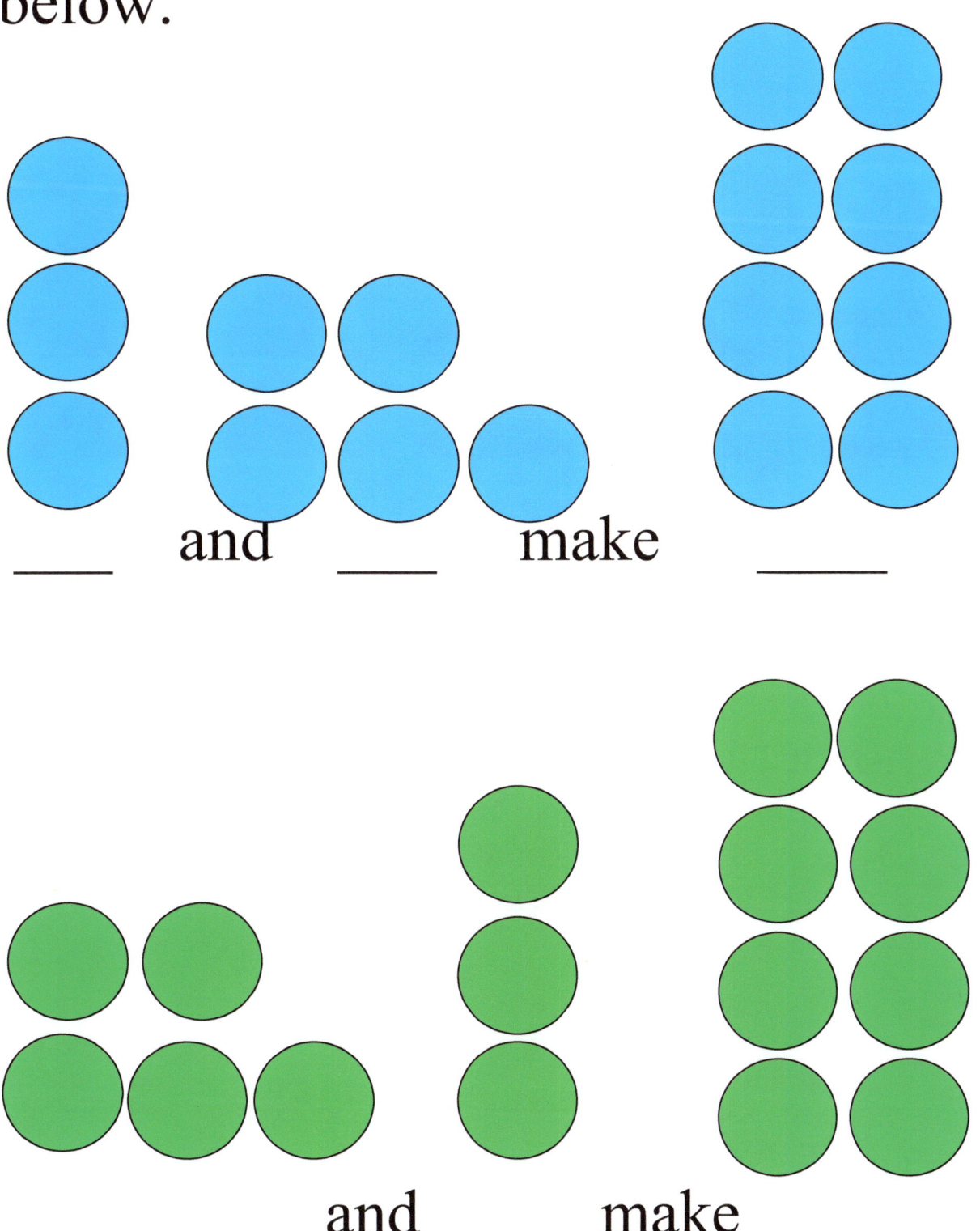

___ and ___ make ___

___ and ___ make ___

Eight

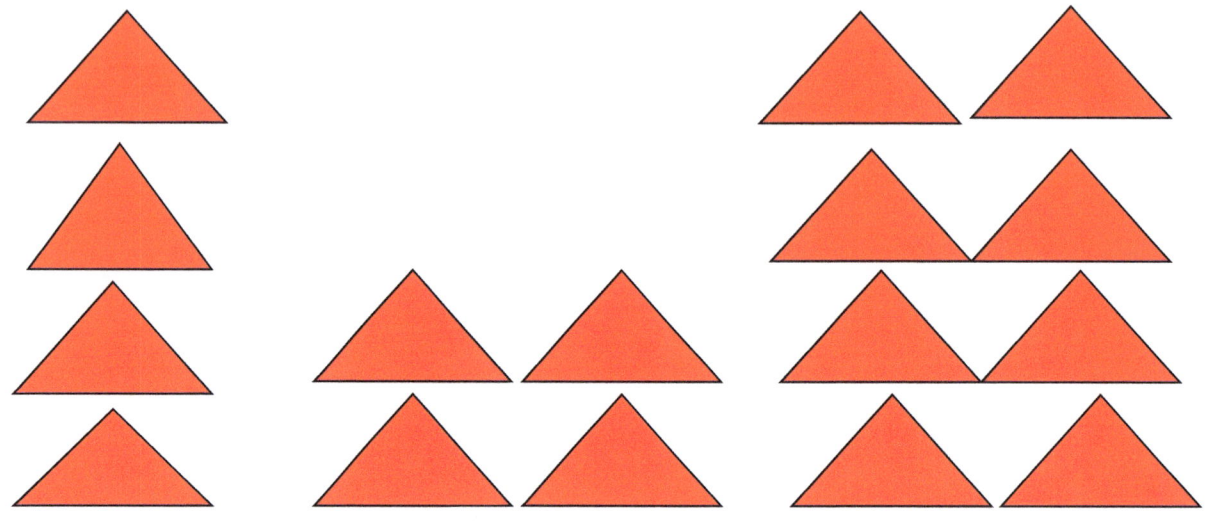

4 and 4 make 8

Draw triangles on the lines to match the numerals below.

_____ _____ _____

4 and 4 make 8

Count the triangles then fill in the blanks.

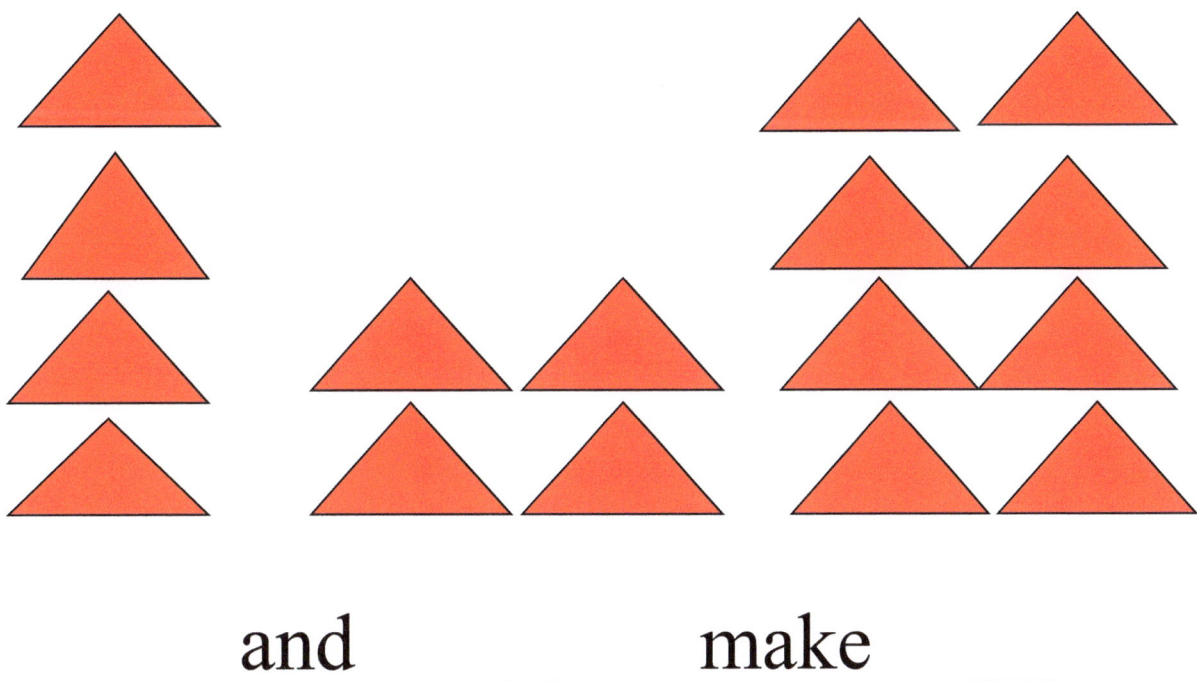

___ and ___ make ___

Draw lines to match the numerals to the correct number of shape or shapes.

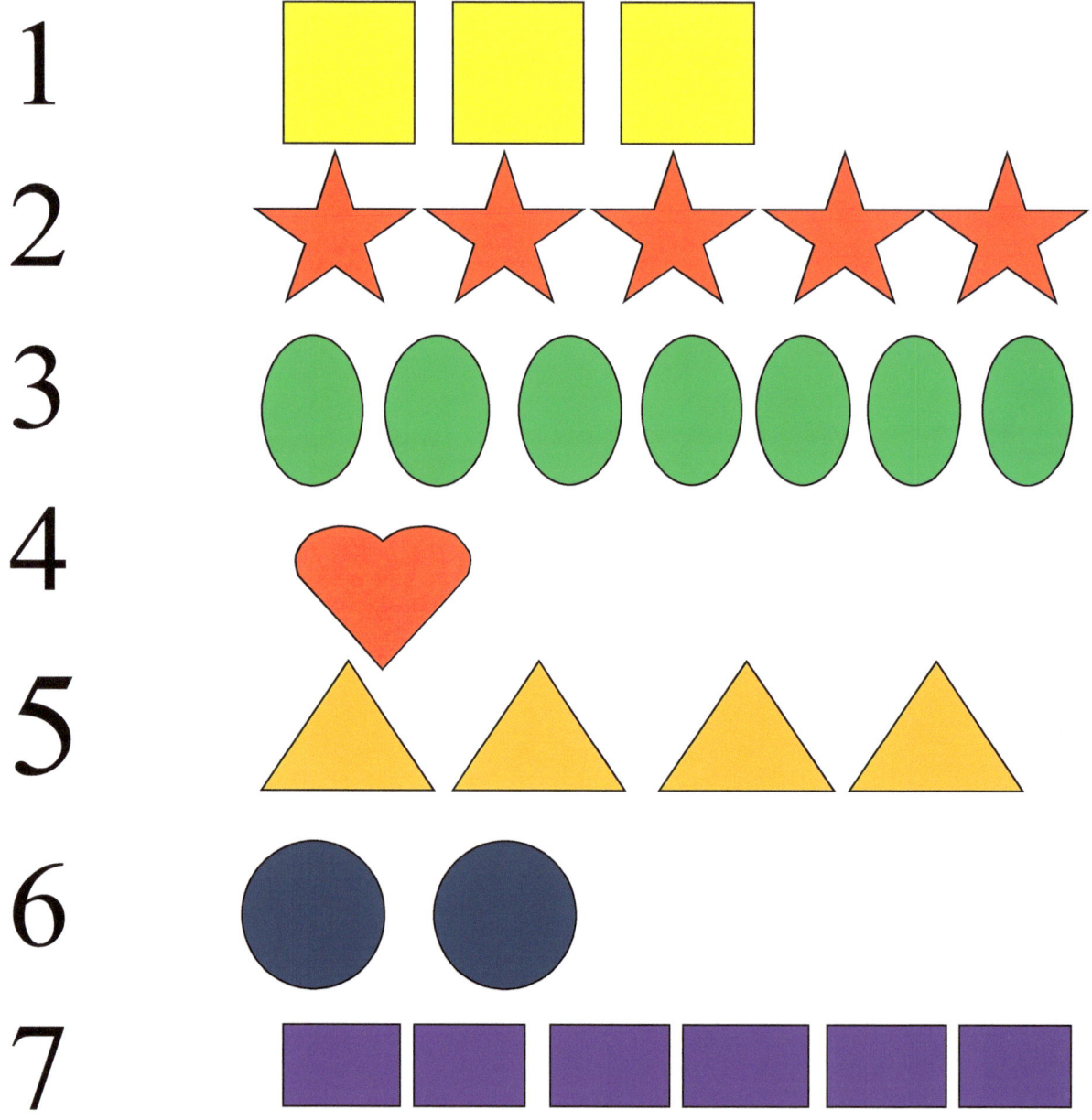

Draw shapes on the lines to match the numerals.

3

4

5

6

7

8

Colouring

1. Colour 4 diamonds blue.

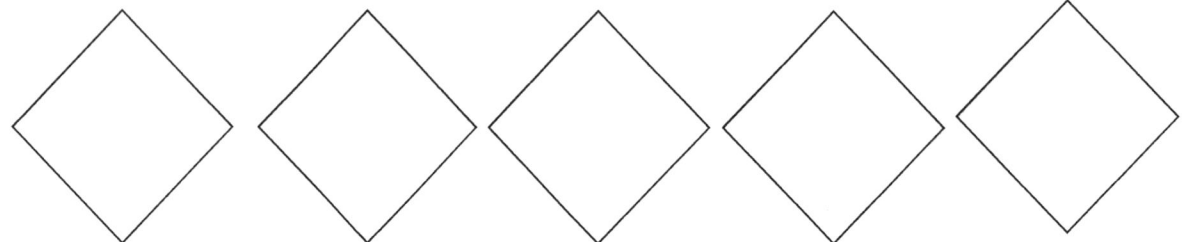

2. Colour 3 squares green.

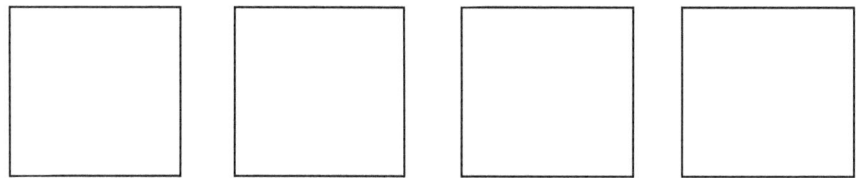

3. Colour 8 ovals purple.

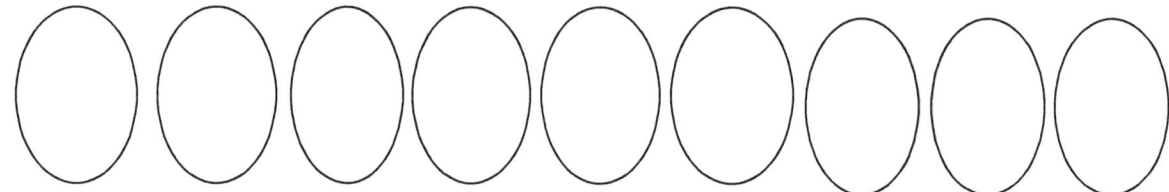

4. Colour 5 stars red.

Circle the numeral which comes next.

3   4   5   3   4 5

1   2   3   1   2 5

5   6   7   5   7 6

6   7   8   6   7 8

1   2   3   1   3 2

Draw and colour 8 diamonds.

Numeral 9

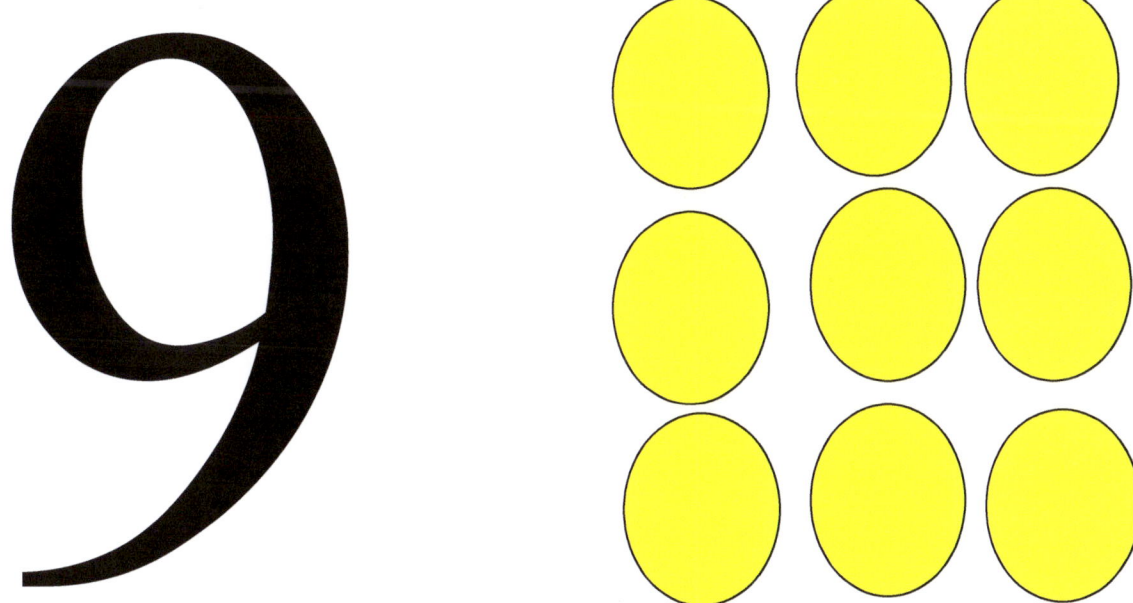

Nine yellow ovals

Nine

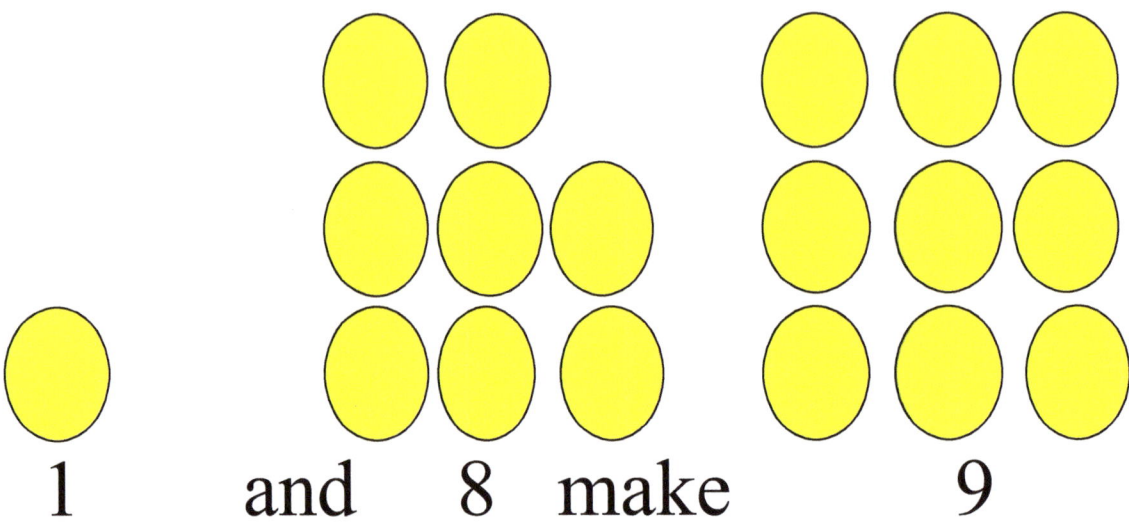

1 and 8 make 9

Draw an ovals or ovals on the lines to match the numerals below.

____ _____ _____
1 and 8 make 9

Nine

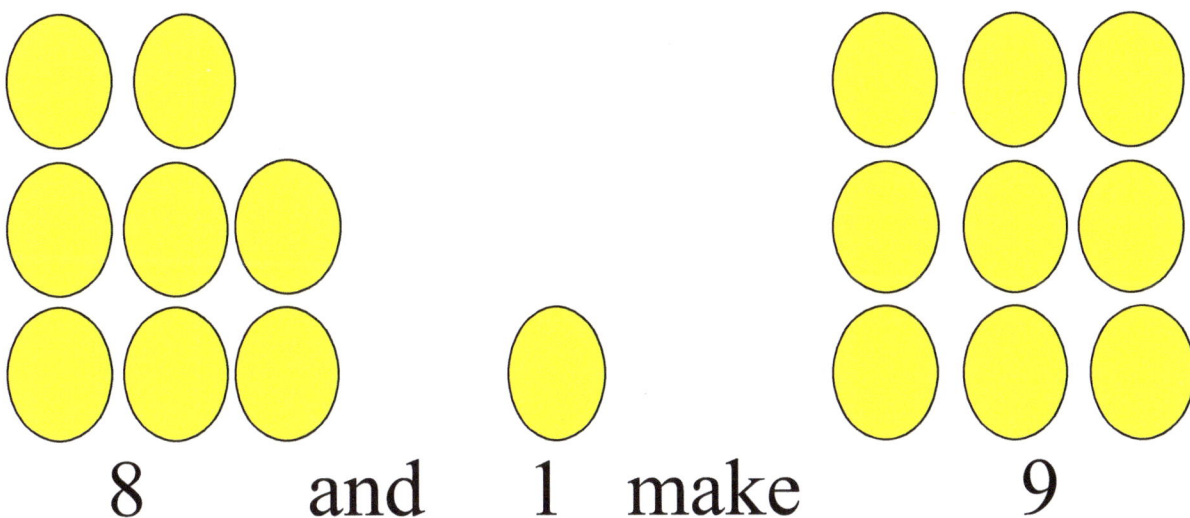

8 and 1 make 9

Draw an oval or ovals on the lines to match the numerals below.

_____ _____ _____
8 and 1 make 9

Count the ovals then fill in the blanks.

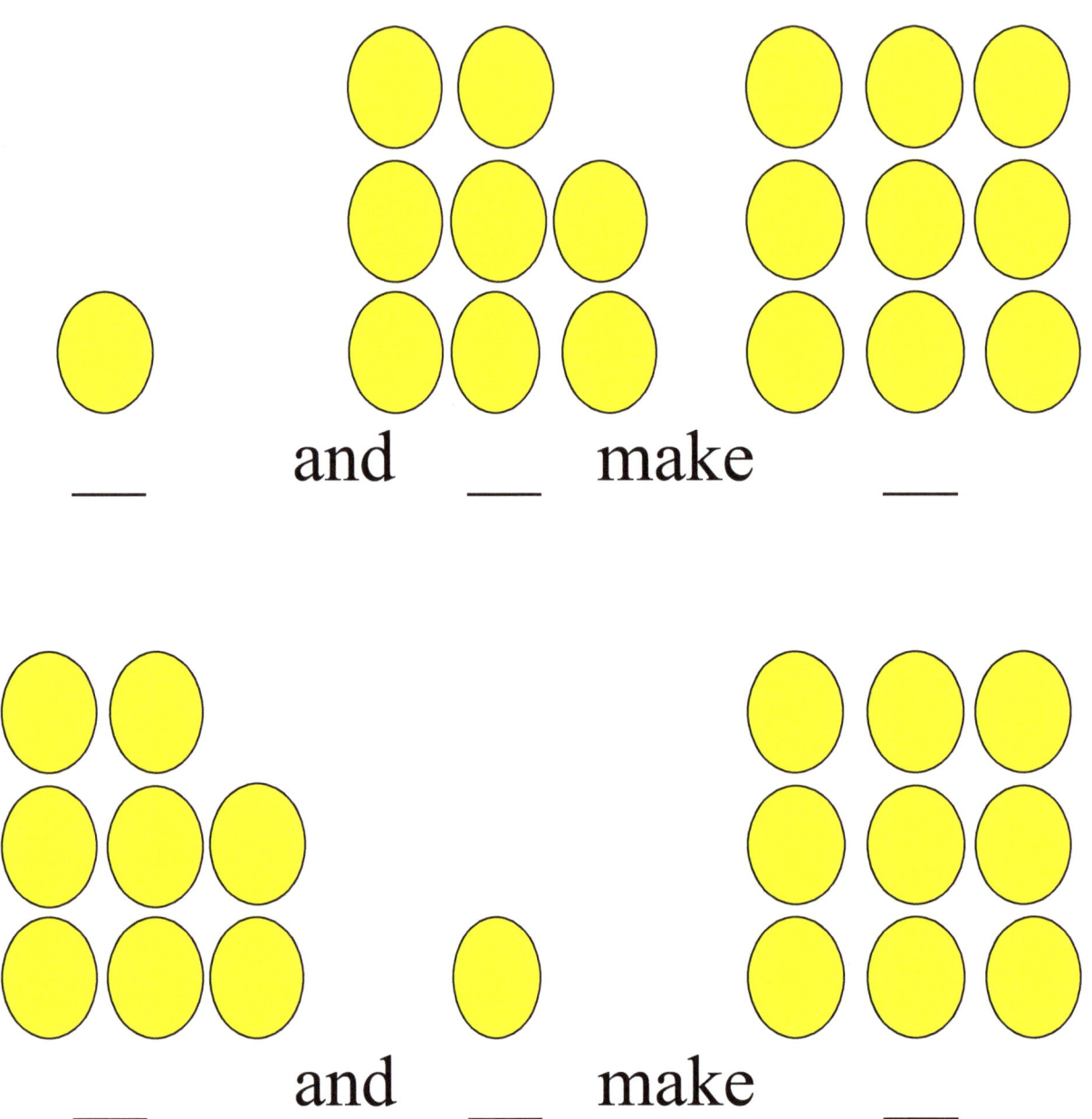

___ and ___ make ___

___ and ___ make ___

Numeral 9

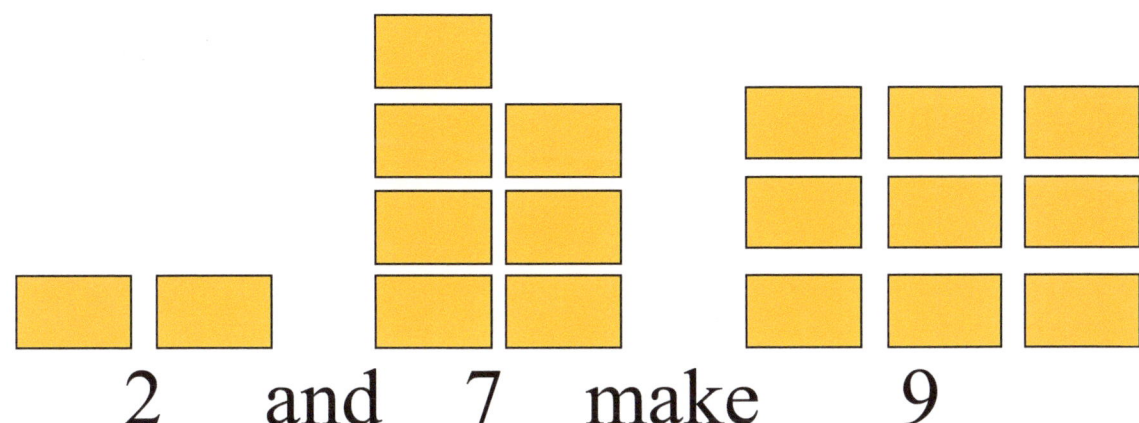

2   and   7   make   9

Draw rectangles on the lines below to match the numerals.

_____   _____   _____

2   and   7   make   9

Numeral 9

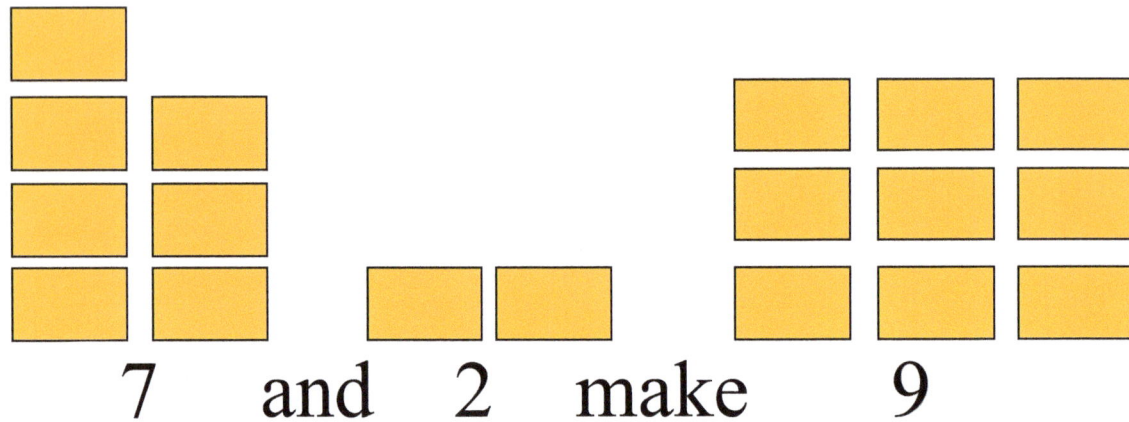

7 and 2 make 9

Draw rectangles on the lines below to match the numerals.

_____   _____   _____

7   and   2   make   9

Count the rectangles then fill in the blanks.

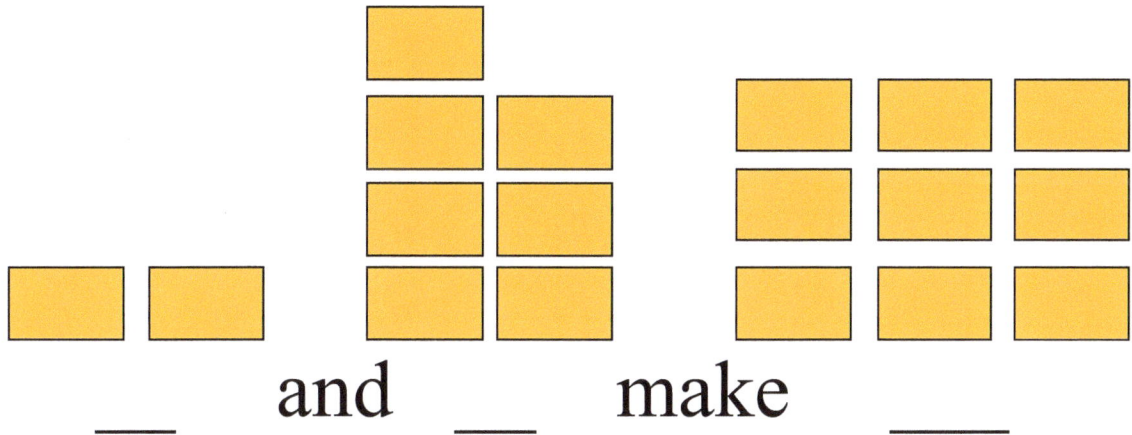

__ and __ make ___

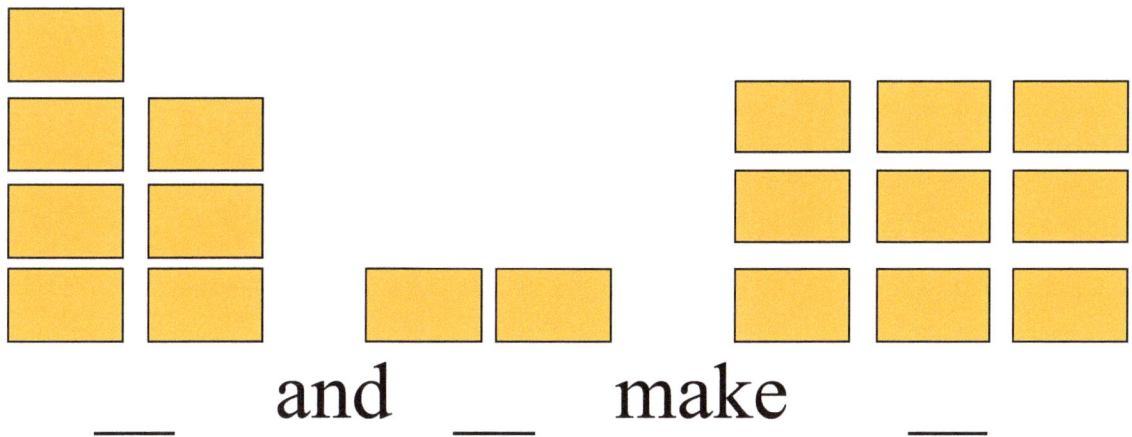

__ and __ make __

Numeral 9

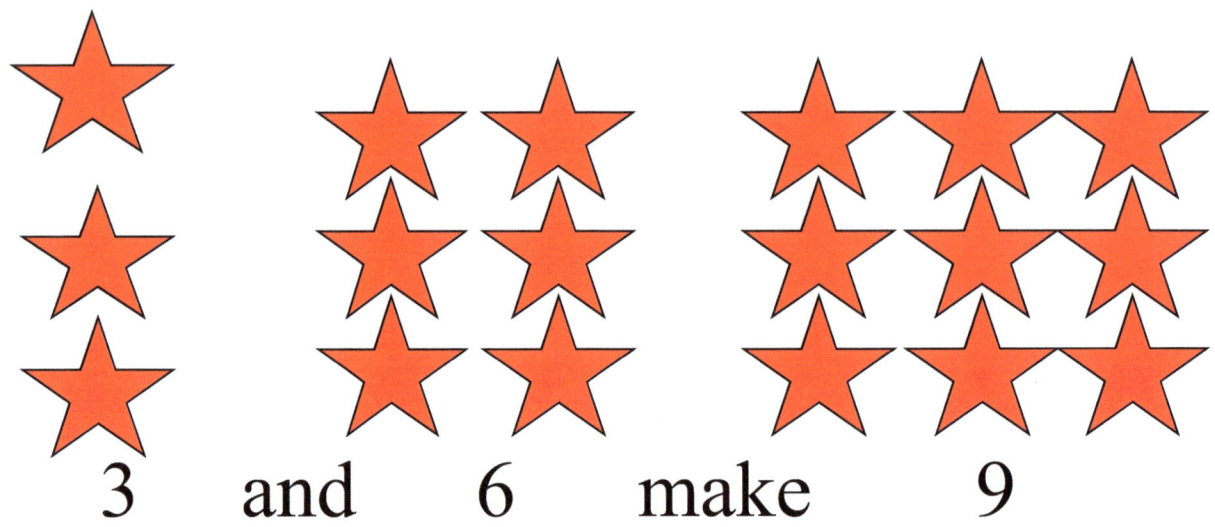

3 and 6 make 9

Draw stars on the lines below to match the numerals.

_____ _____ _____

3 and 6 make 9

Numeral 9

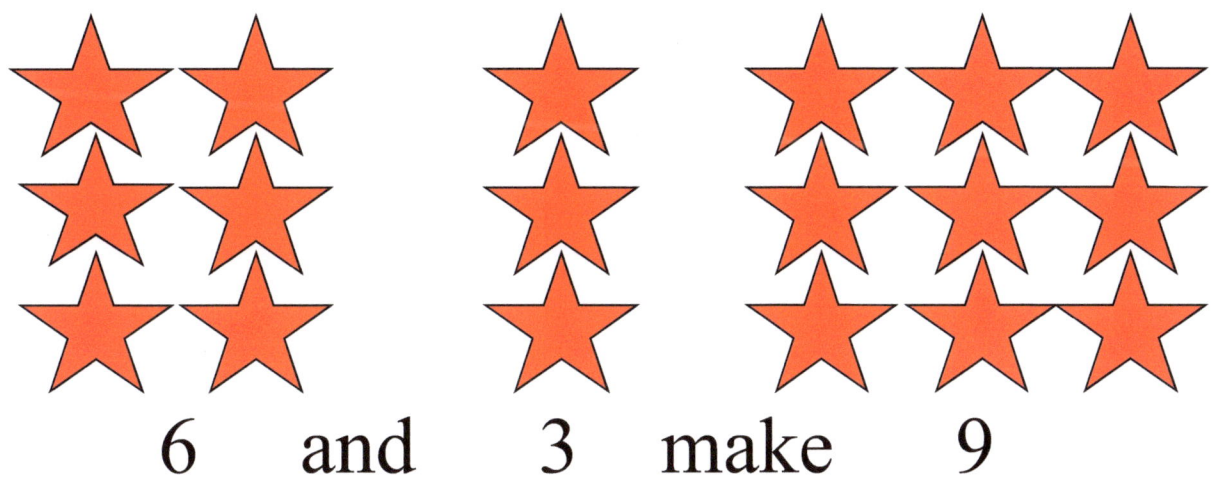

6 and 3 make 9

Draw stars on the lines below to match the numerals.

_____ _____ _____

6 and 3 make 9

Count the stars then fill in the blanks.

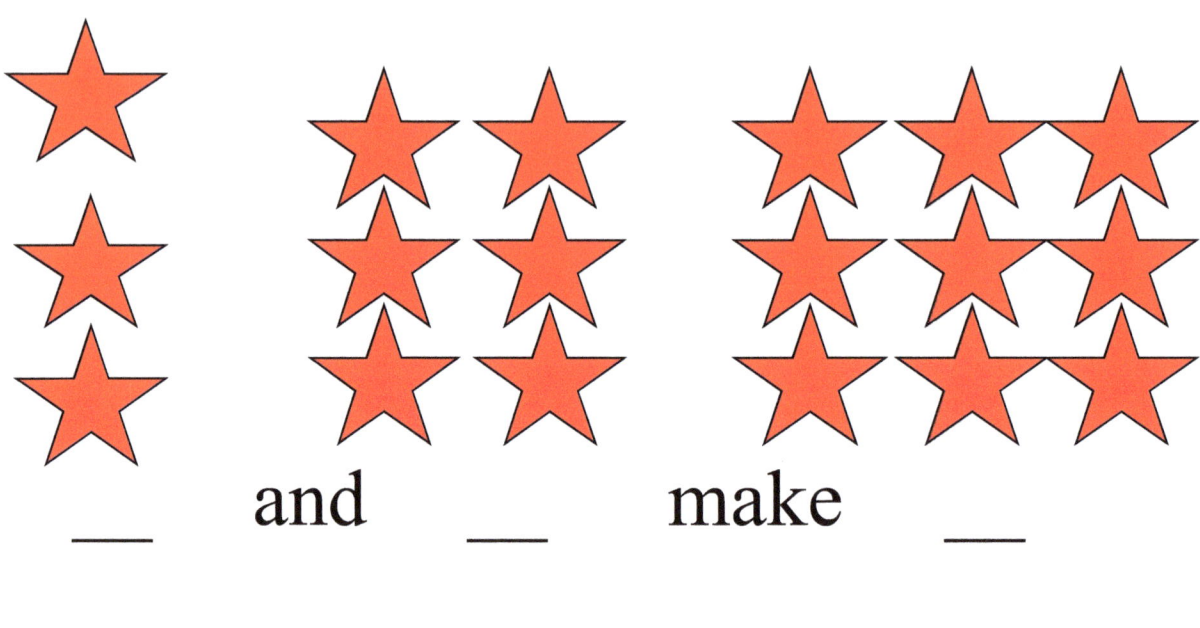

__ and __ make __

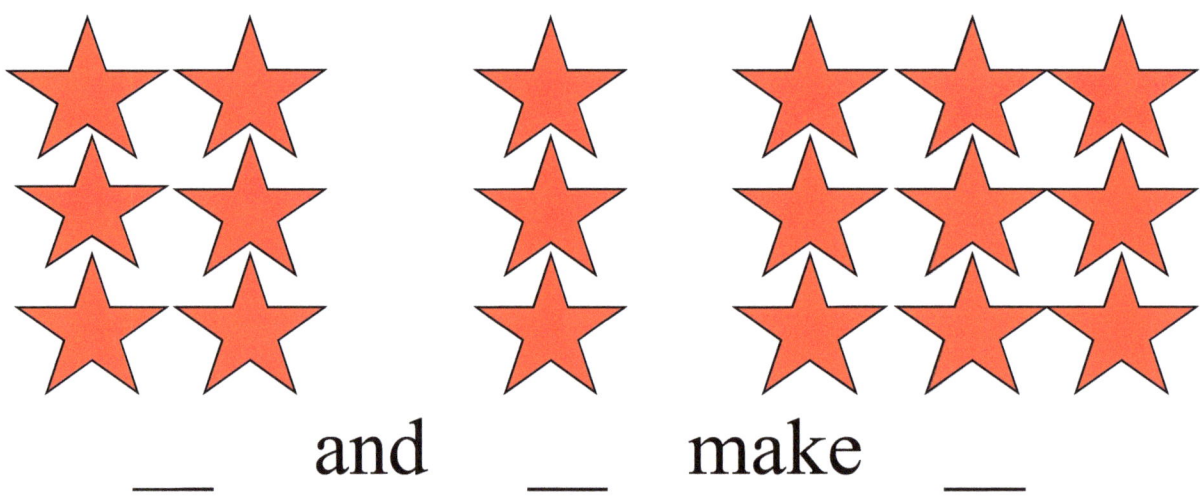

__ and __ make __

Numeral 9

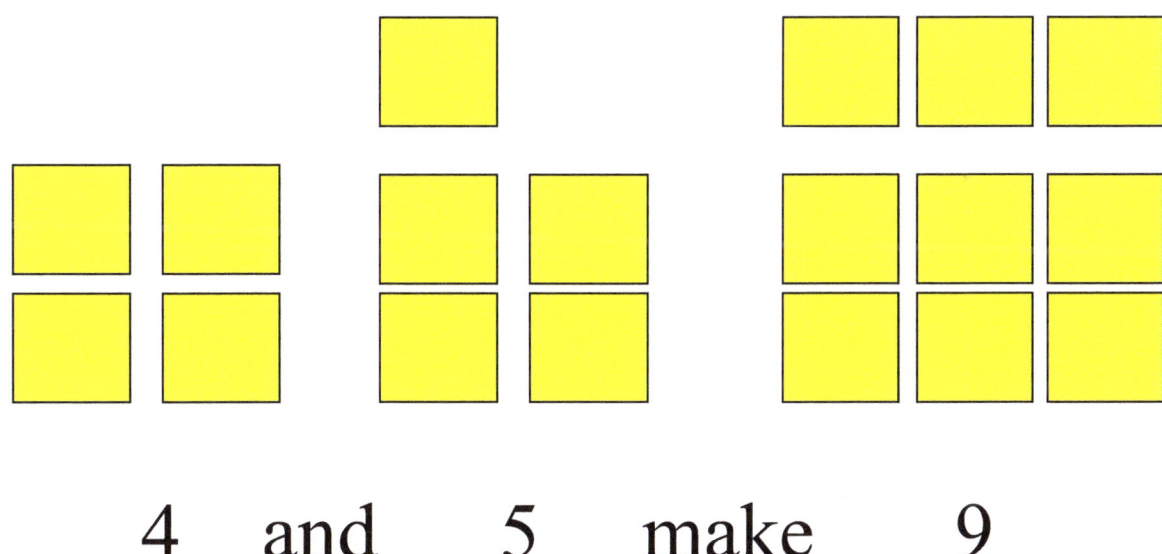

4 and 5 make 9

Draw squares on the lines to match the numerals below.

_____ _____ _____

4 and 5 make 9

Numeral nine

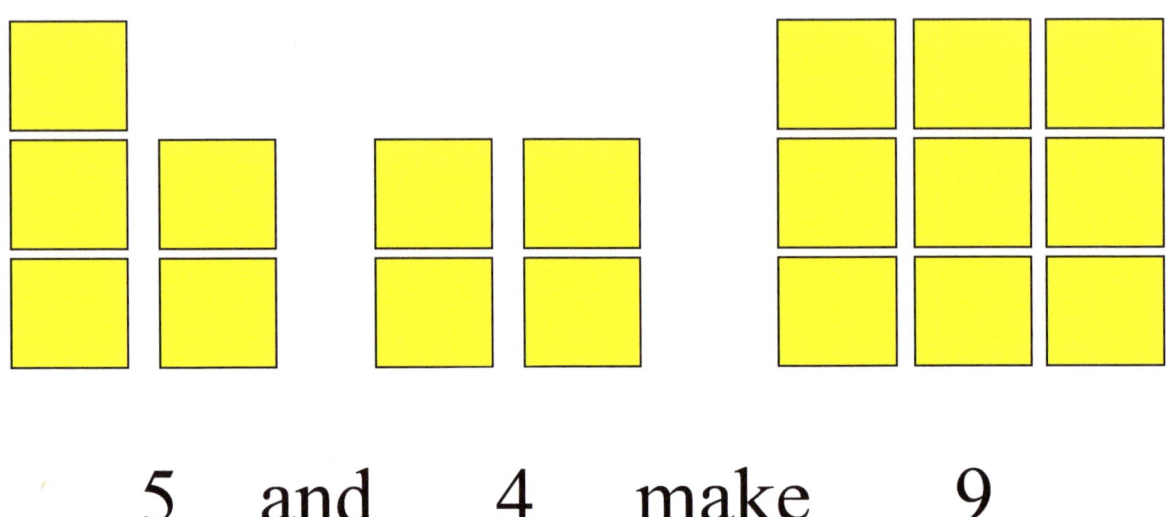

5 and 4 make 9

Draw squares on the lines to match the numerals below.

_____   _____   _____

5 and 4 make 9

Count the squares then fill in the blanks.

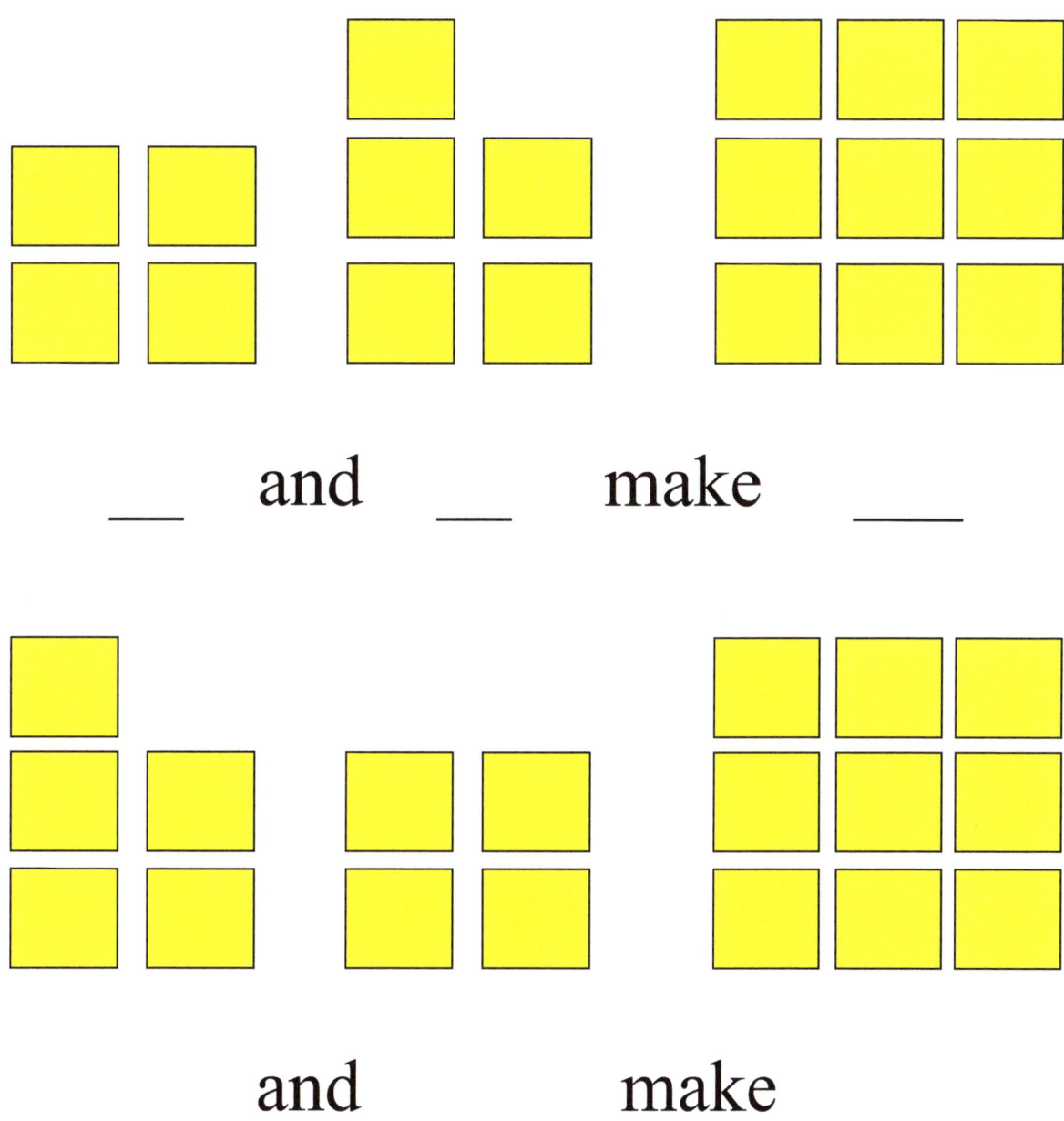

__ and __ make __

__ and __ make __

Numerals

Draw lines to match the numerals to the correct number of shape or shapes.

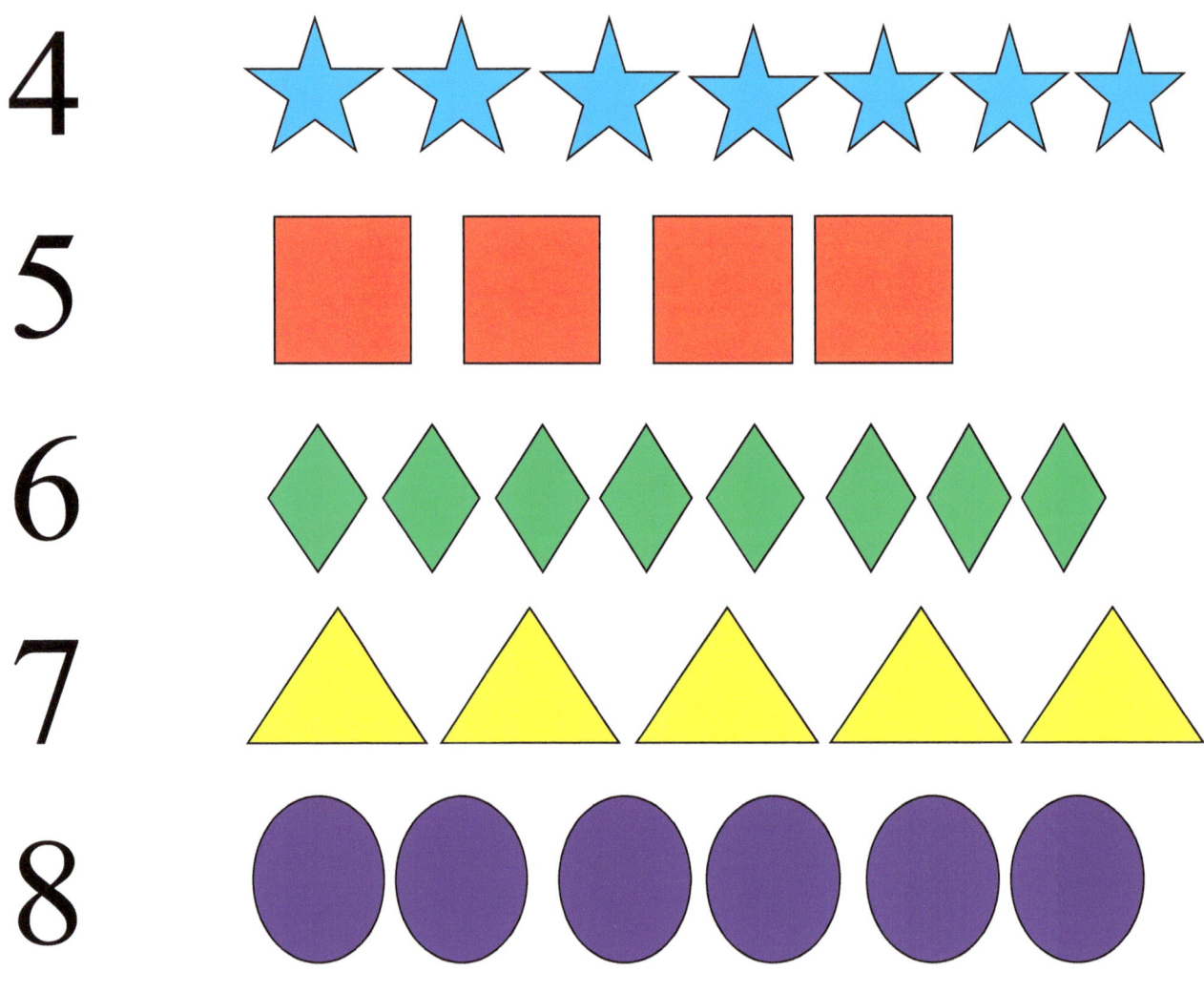

Draw shapes on the lines to match the numerals.

7 _____

8 _____

9 _____

Colouring

1. Colour 3 triangles yellow purple.

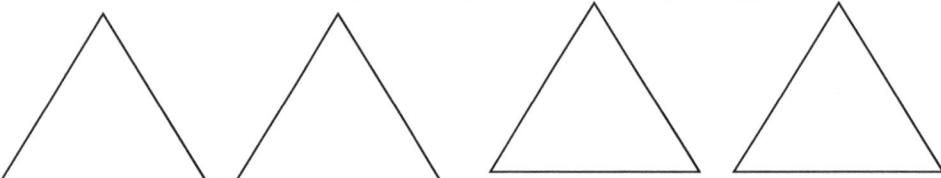

2. Colour 4 diamonds red.

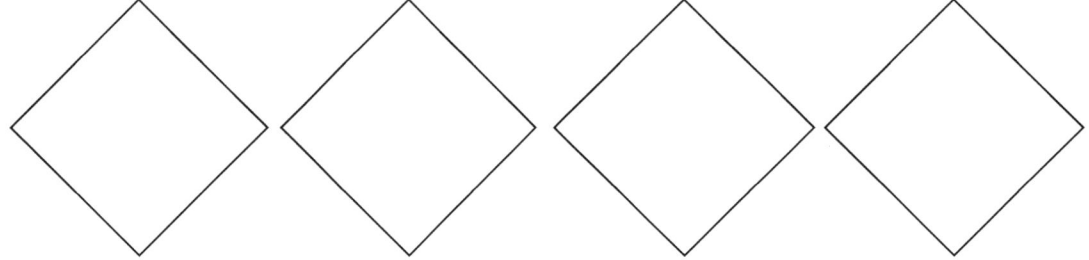

3. Colour 2 squares green.

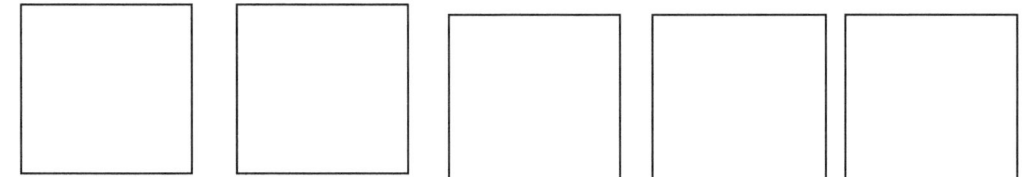

4. Colour 5 stars blue.

Shapes Patterns
Draw and colour the shape that comes next.

Draw and colour 9 ovals.

Numeral 10

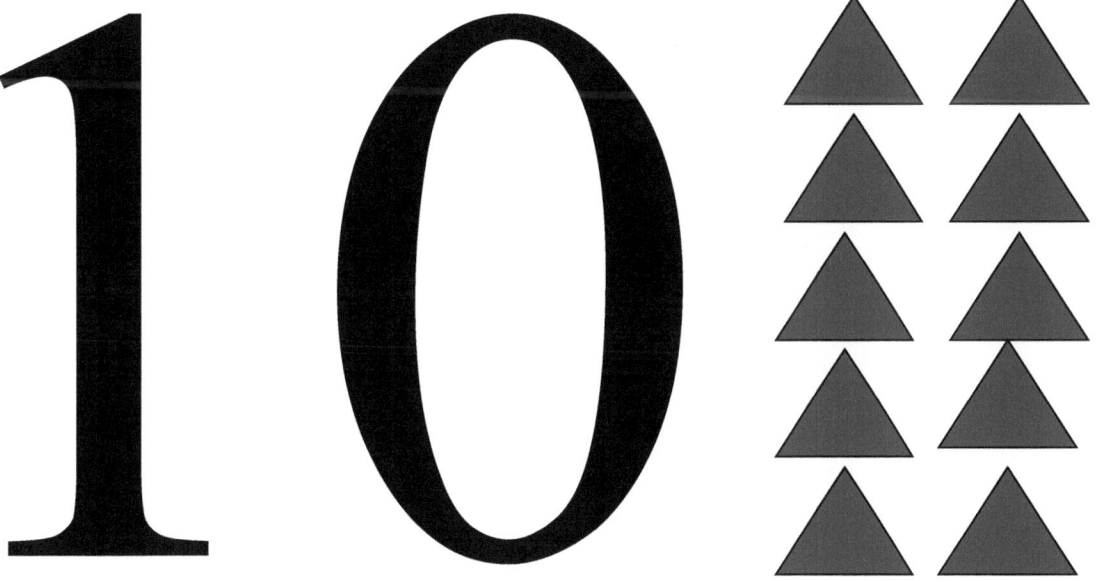

Ten blue triangles

Numeral 10

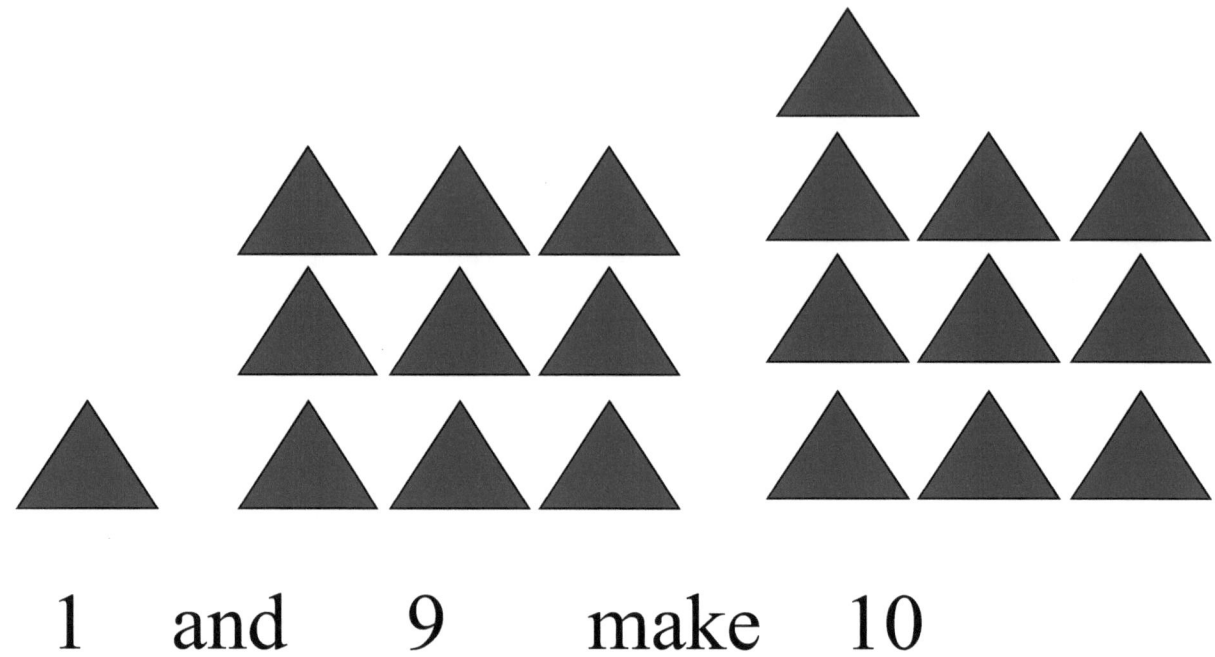

1 and 9 make 10

Draw a triangle or triangles below to match the numerals.

1 and 9 make 10

Numeral 10

9   and   1   make   10

Draw a triangle or triangles below to match the numerals.

9   and   1   make   10

Count the triangles then fill in the blanks.

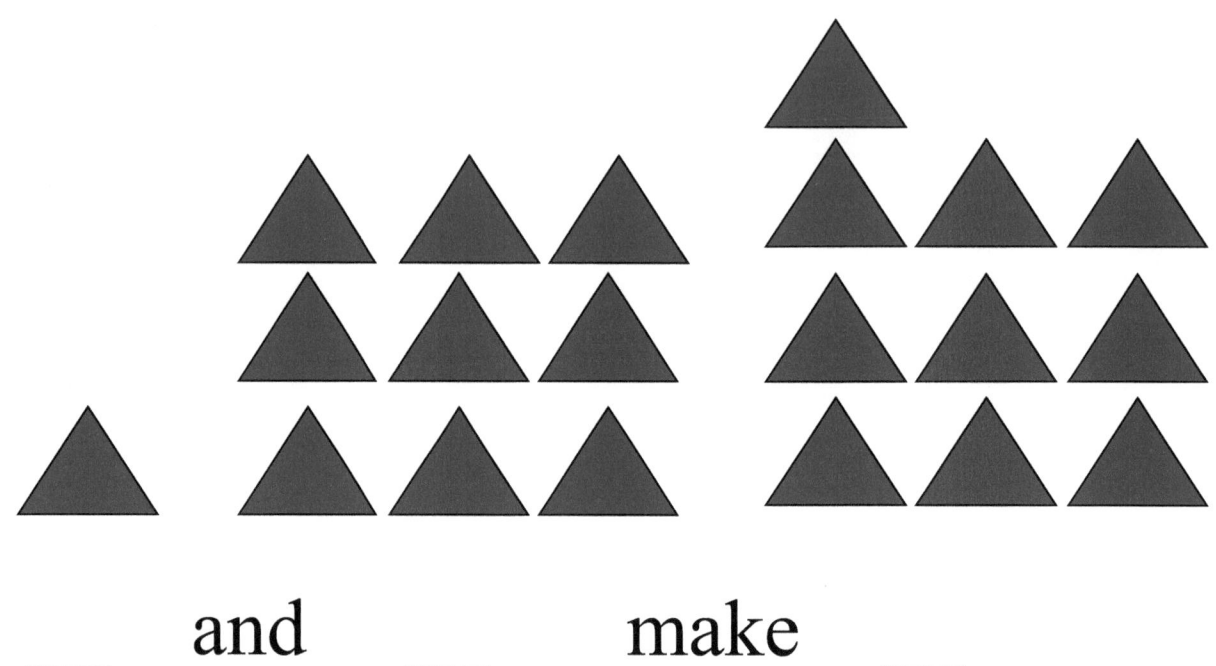

__ and __ make __

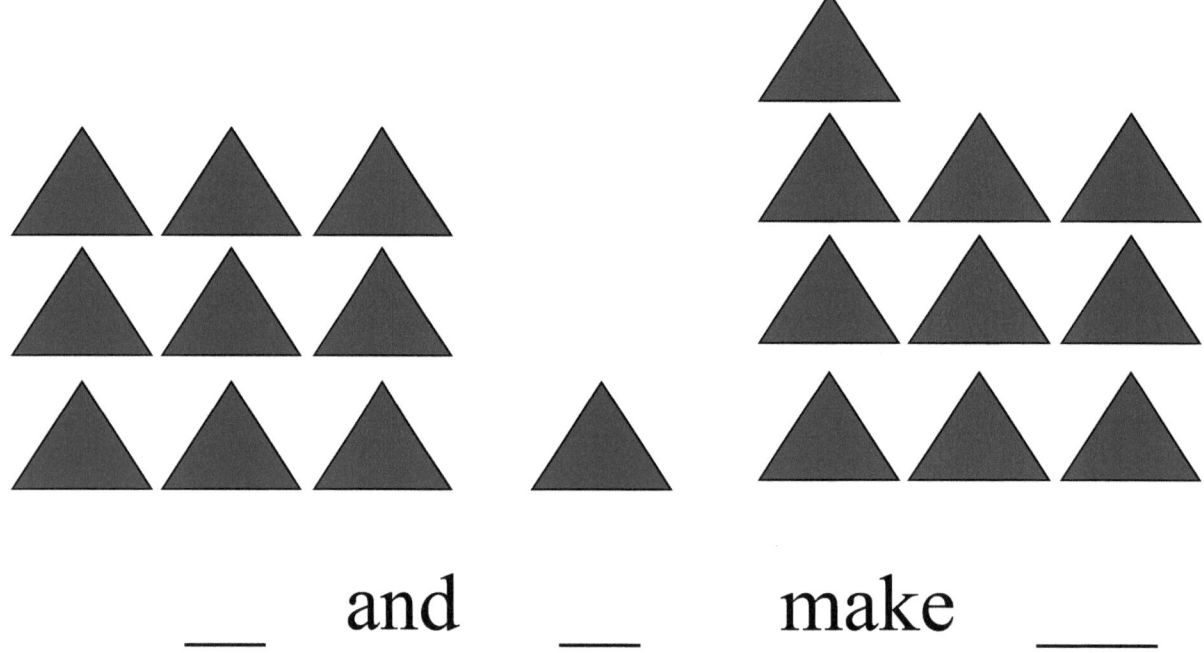

__ and __ make ___

Numeral 10

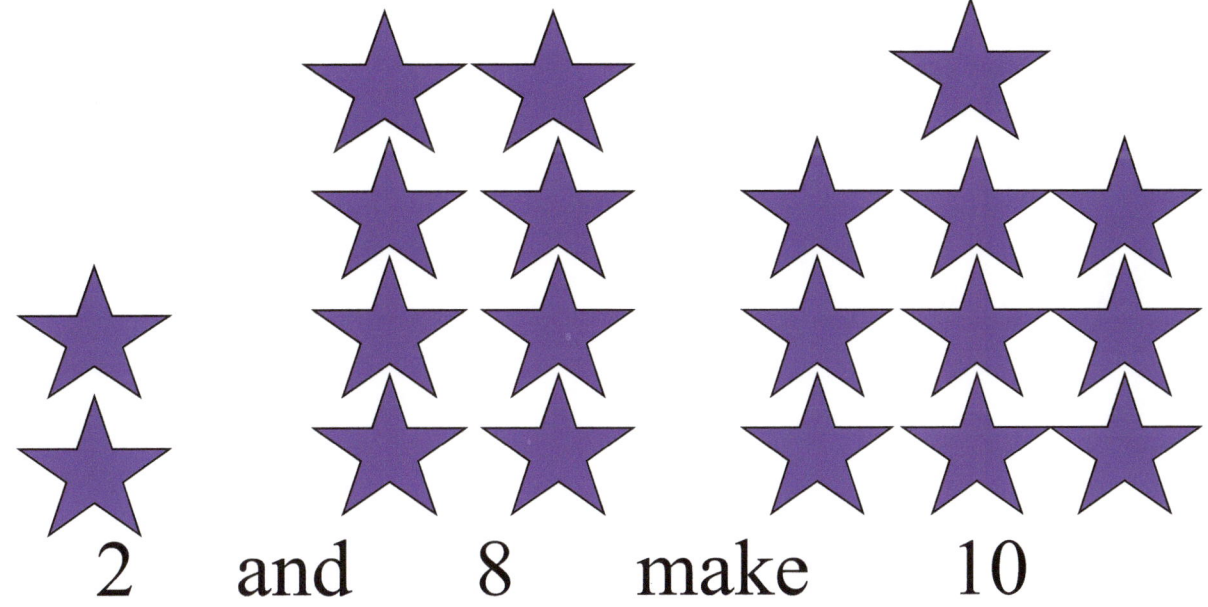

2 and 8 make 10

Draw stars on the lines below to match the numerals.

_____ _____ _____
   2   and   8   make   10

Numeral 10

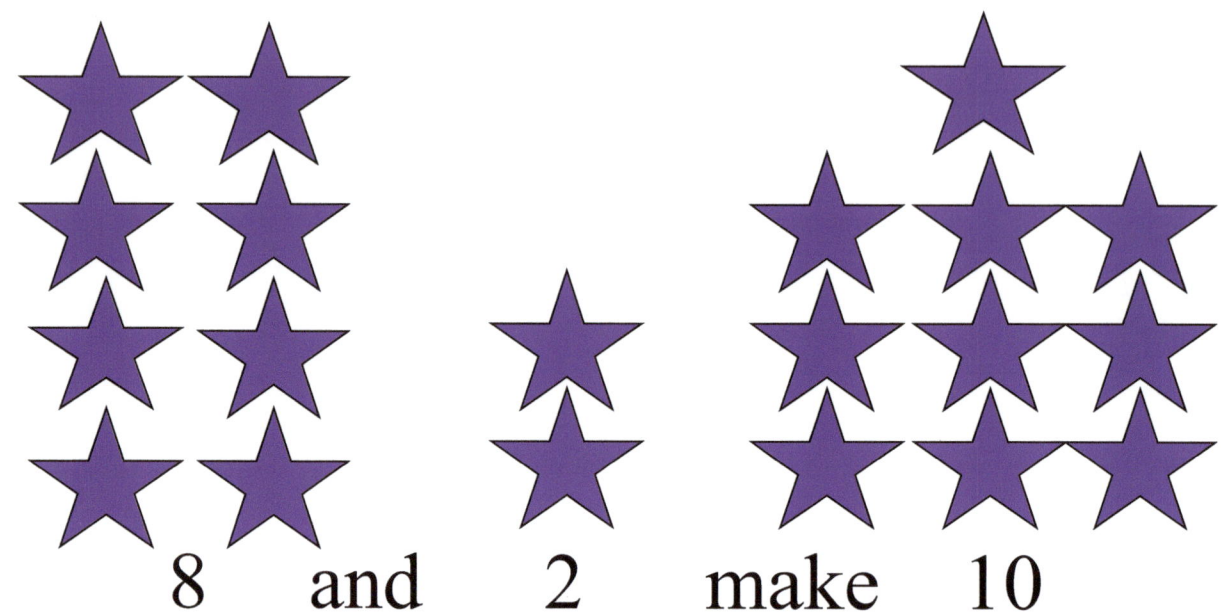

8 and 2 make 10

Draw stars on the lines below to match the numerals.

_____   _____   _____
     8   and   2   make   10

Count the stars then fill in the blanks.

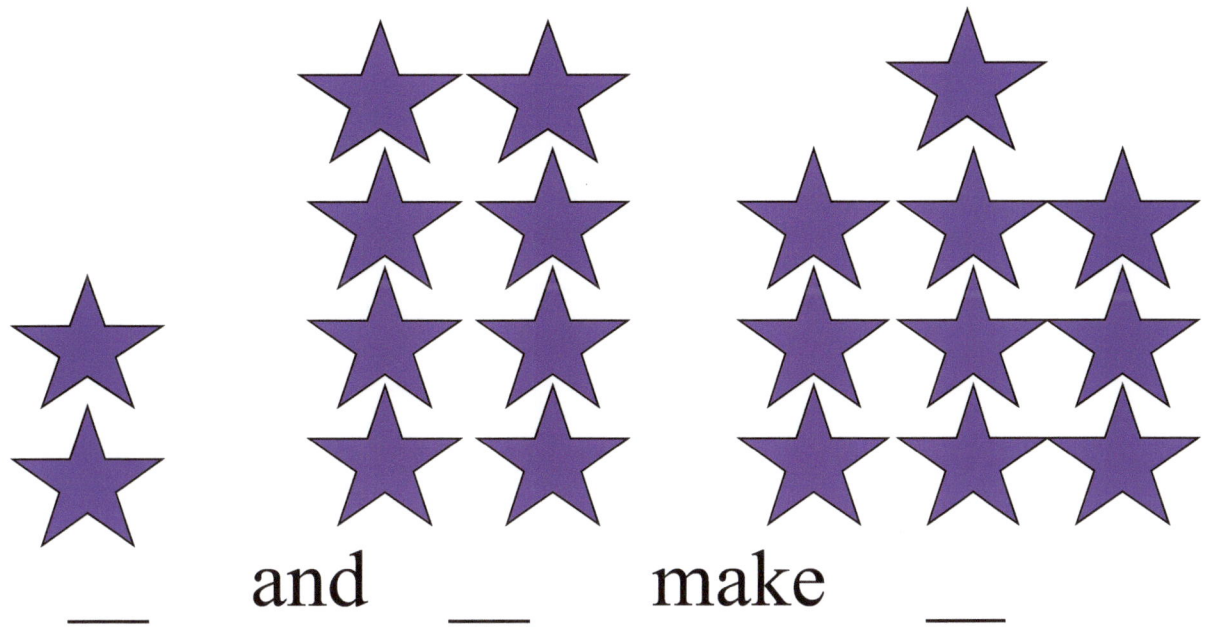

__ and __ make __

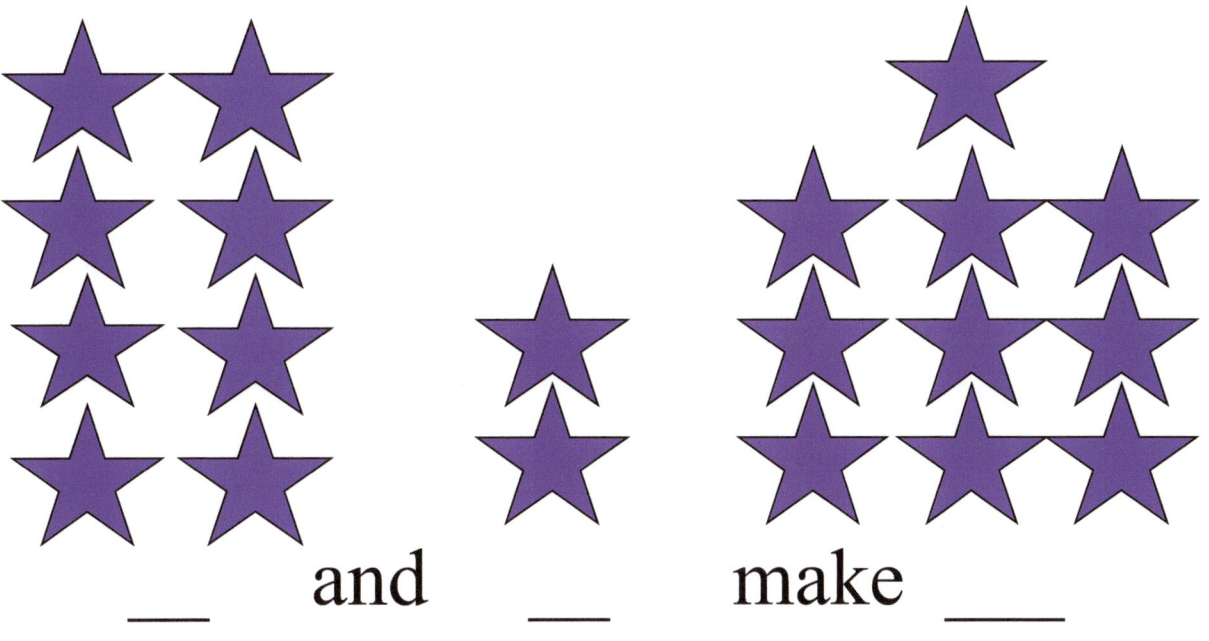

__ and __ make __

Numeral 10

3 and 7 make 10

Draw hearts to match the numerals below.

3 and 7 make 10

Numeral 10

7 and 3 make 10

Draw hearts to match the numerals below.

7 and 3 make 10

Count the hearts then fill in the blanks.

___ and ___ make ___

___ and ___ make ___

Ten

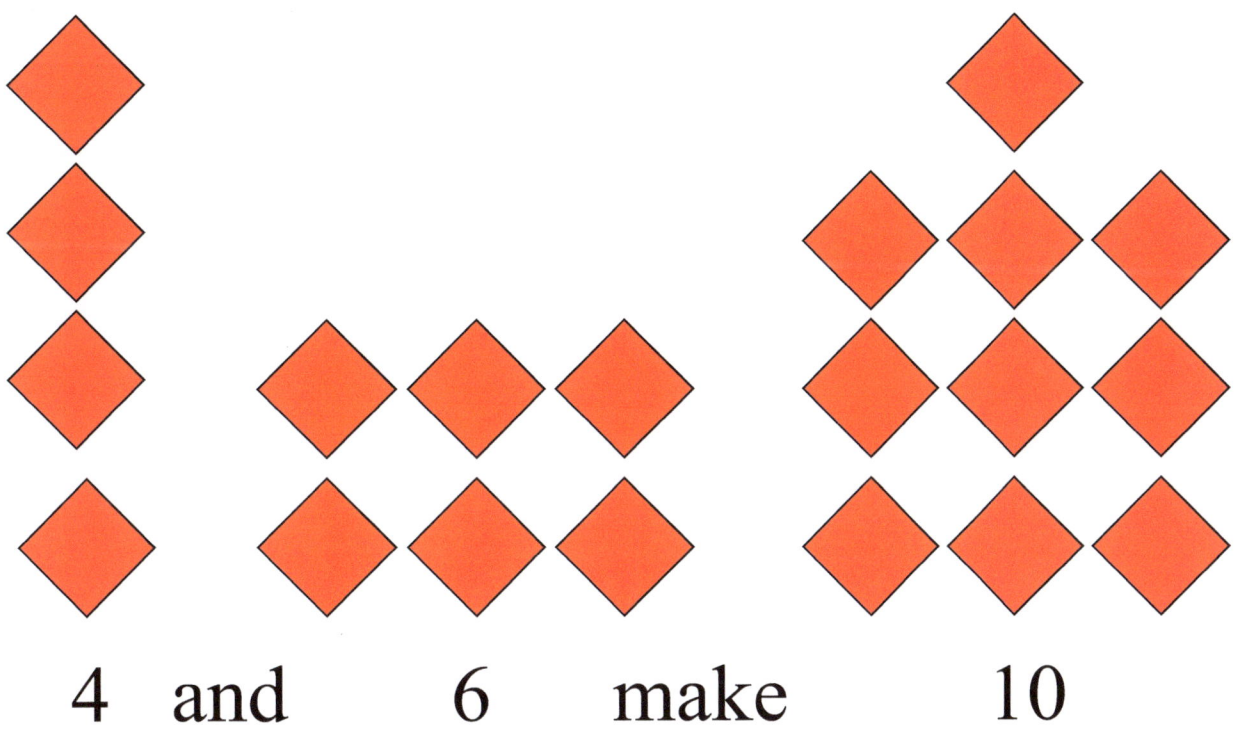

4 and 6 make 10

Draw diamonds to match the numerals below.

4 and 6 make 10

Ten

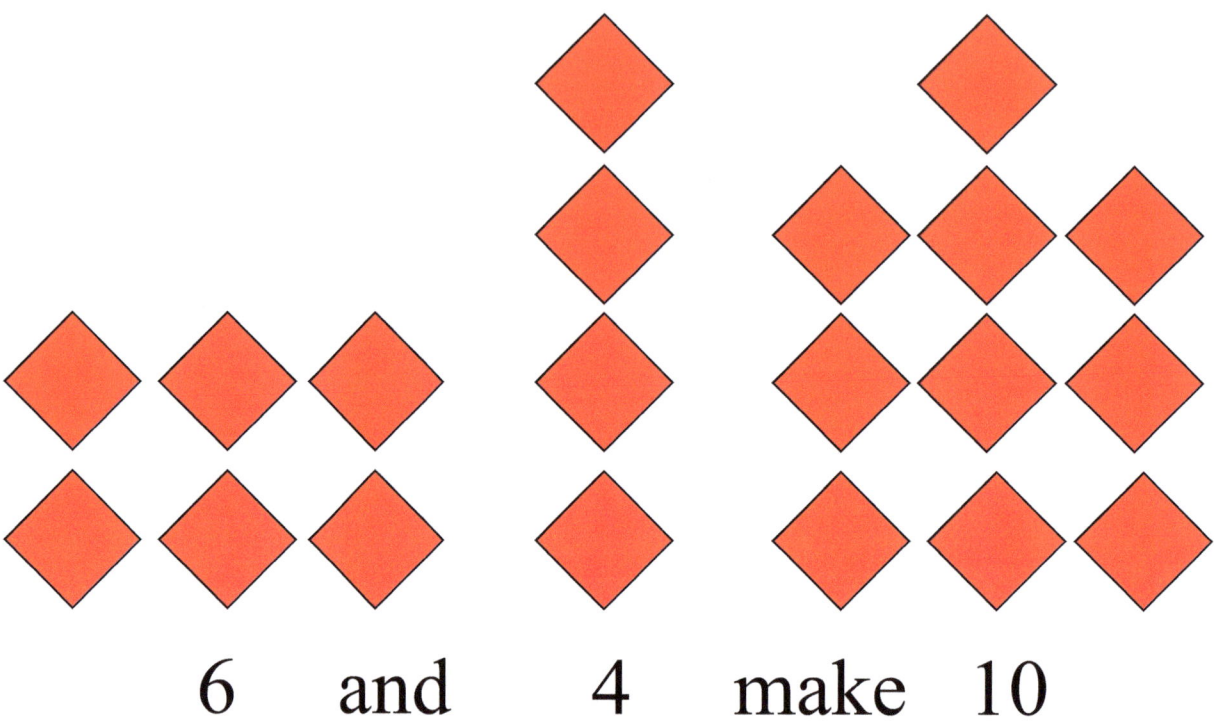

6 and 4 make 10

Draw diamonds to match the numerals below.

6 and 4 make 10

Count the diamonds then fill in the blanks.

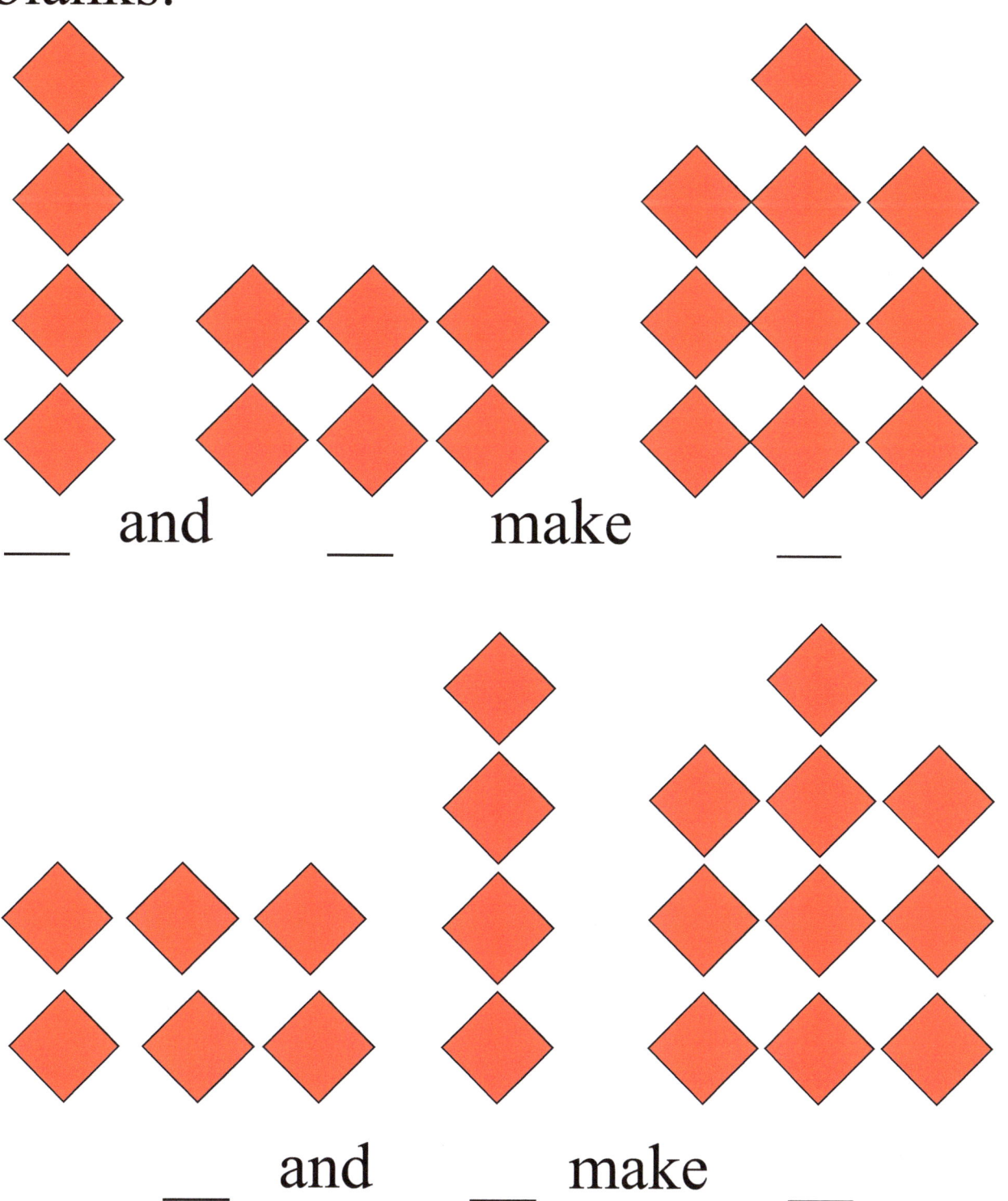

__ and __ make __

__ and __ make __

Numeral 10

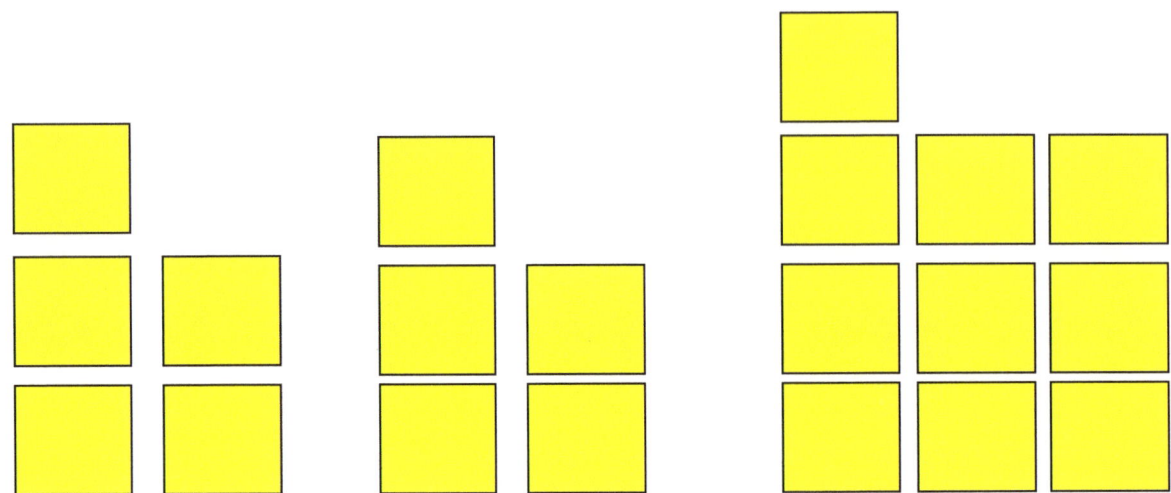

5 and 5 make 10

Draw squares on the lines to match the numerals below.

_____   _____   _____

5 and 5 make 10

Count the squares then fill in the blanks.

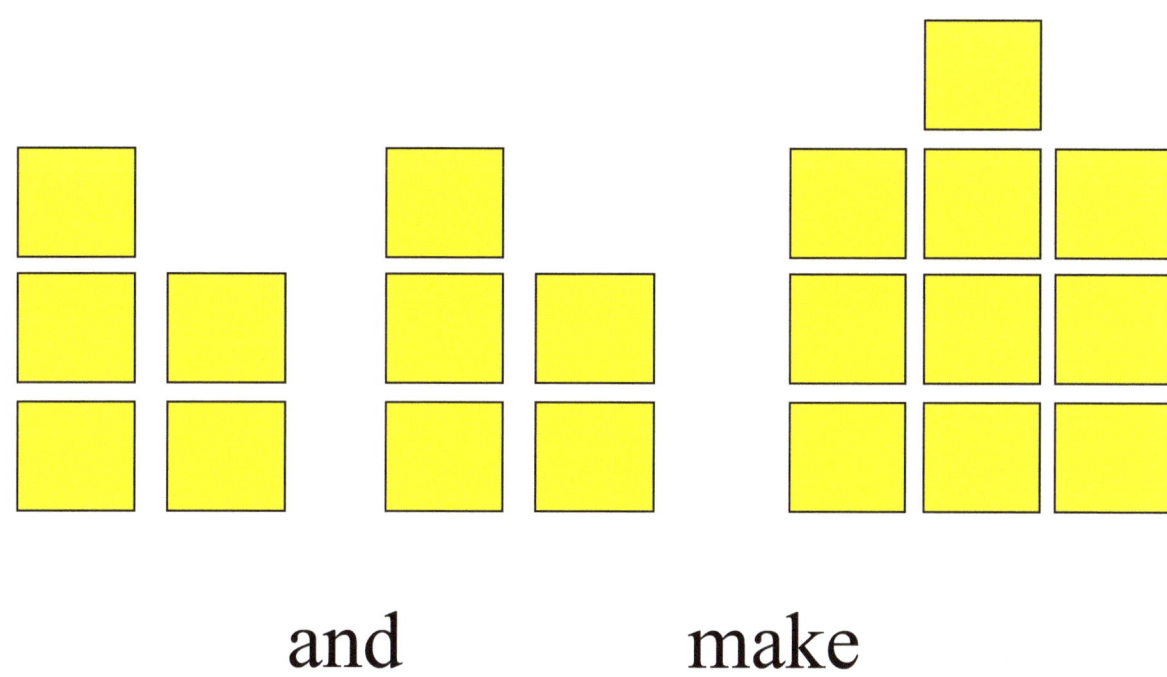

___ and ___ make ___

Colouring

Colour 3 ovals

Colour 3 ovals

Colour 4 ovals

Shapes Patterns
Colour the shape which comes next.

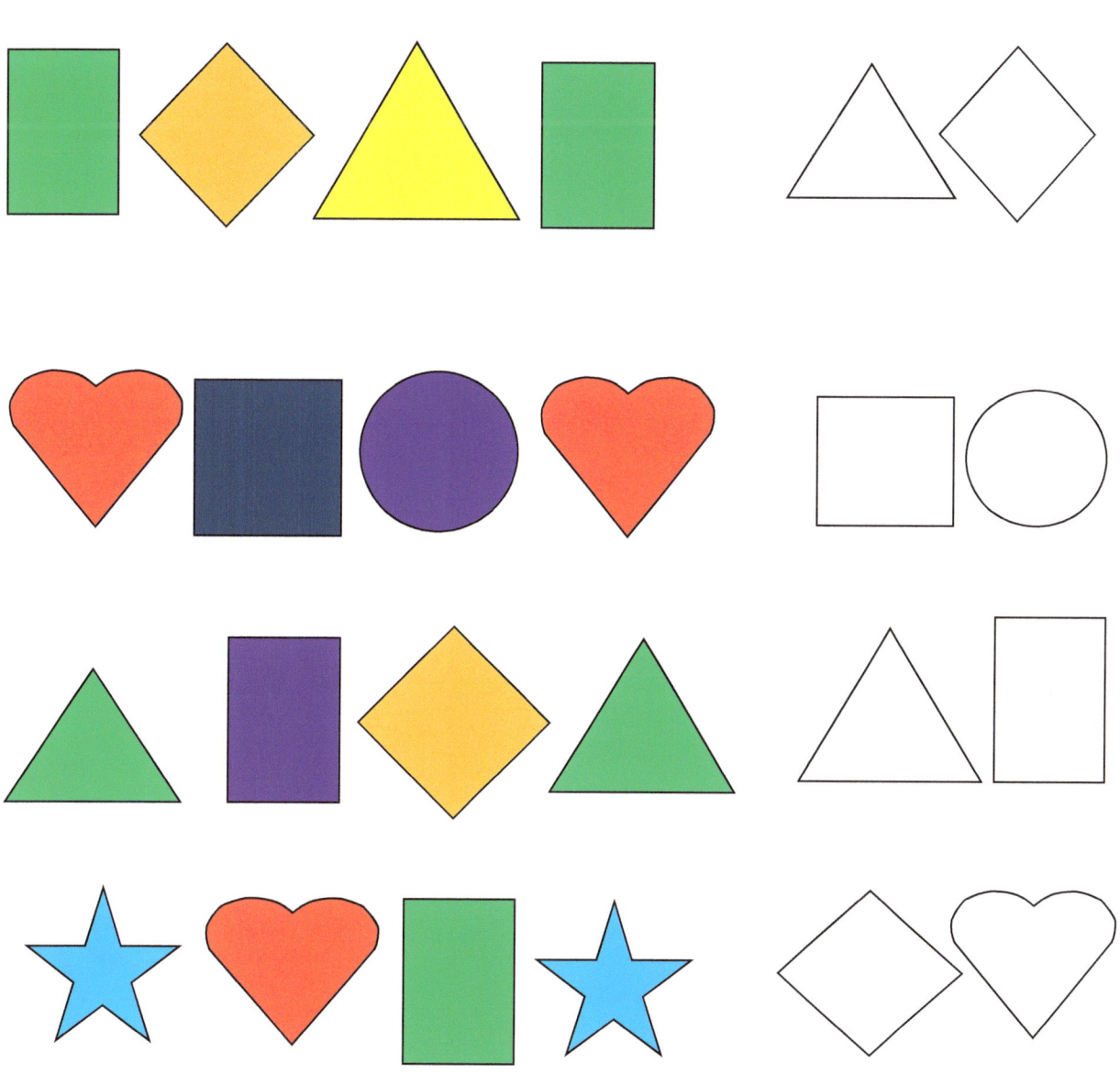

Count the shapes then circle the numeral which tell how many.

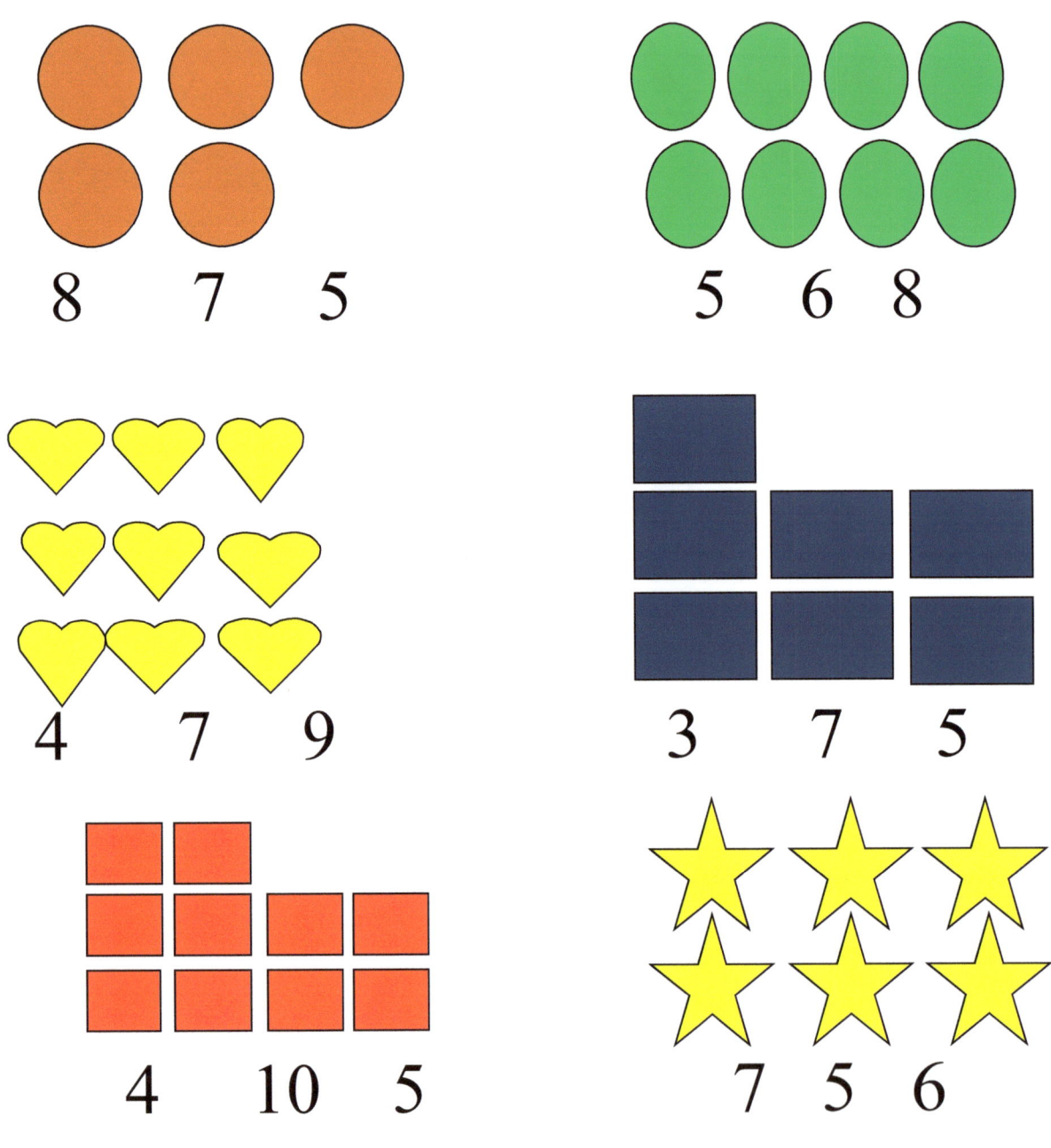

Colouring

1. Colour 8 triangles yellow.

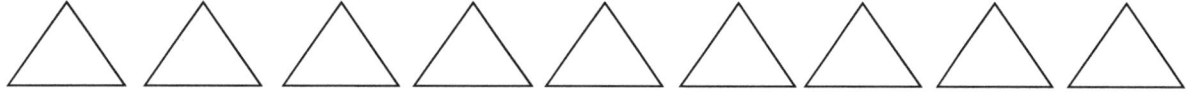

2. Colour 6 diamonds blue

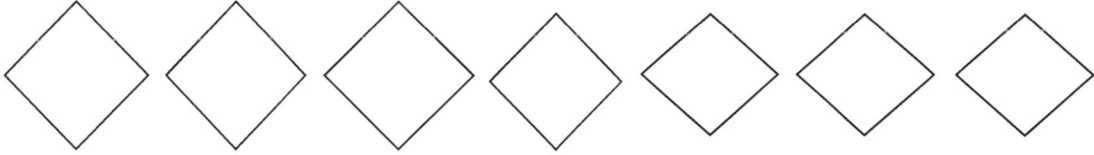

3. Colour 4 rectangles green.

4. Colour 10 ovals orange.

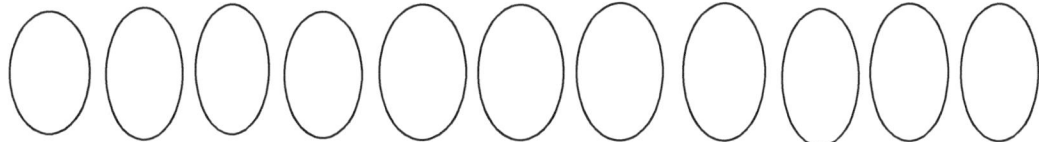

5. Colour 5 stars red.

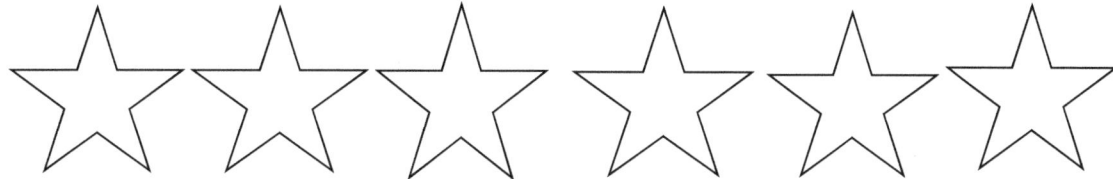

6. Colour 3 hearts yellow.

Numerals 1 - 10

Write the missing numerals.

1 ___   3 ___

5 ___   7 ___

9 ___

Draw shapes on the lines to match the numerals.

5 _____

6 _____

7 _____

8 _____

9 _____

10 _____

Draw lines to the match the shapes that are the same.

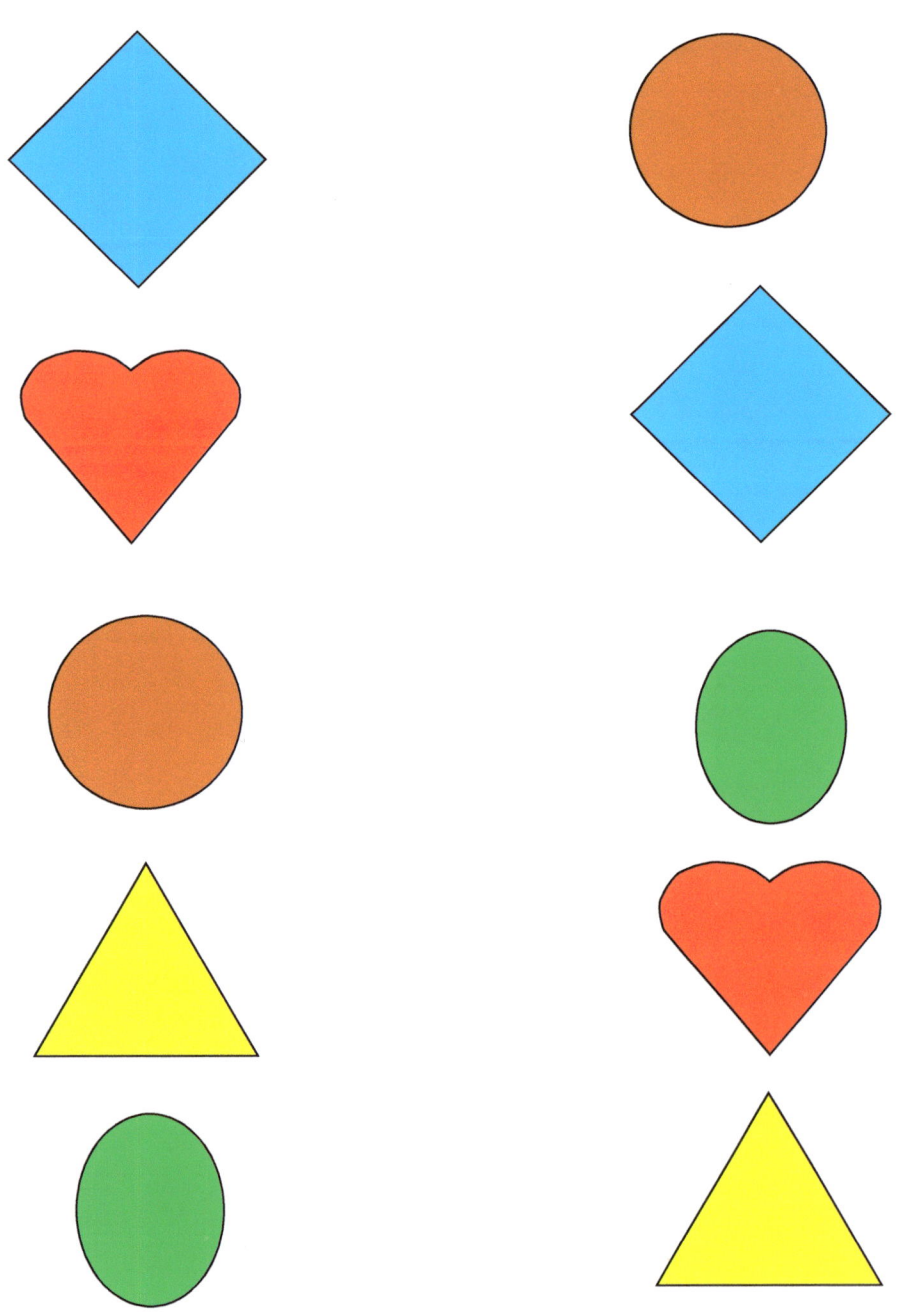

Draw and colour ten triangles.

www.ingramcontent.com/pod-product-compliance
Lightning Source LLC
Chambersburg PA
CBHW051148220526
45473CB00003B/698